PREFATORY NOTE.

The investigations the results of which are herein set forth were carried out by the aid of certain grants from the Carnegie Institution of Washington. The author desires to express his deep sense of obligation for the aid thus rendered. The first five papers were prepared at the Zoological Laboratory of the University of Michigan, and were submitted to the Carnegie Institution for publication August 1, 1903. To the third paper some additions were made in February, 1904. The sixth and seventh papers were prepared at the Naples Zoological Station, while the writer was acting as Research Assistant of the Carnegie Institution, and were transmitted for publication in January and March, respectively, 1904.

LIST OF PAPERS.

1. Reactions to Heat and Cold in the Ciliate Infusoria.
2. Reactions to Light in Ciliates and Flagellates.
3. Reactions to Stimuli in Certain Rotifera.
4. The Theory of Tropisms.
5. Physiological States as Determining Factors in the Behavior of Lower Organisms.
6. The Movements and Reactions of Amœba.
7. The Method of Trial and Error in the Behavior of Lower Organisms.

FIRST PAPER.

REACTIONS TO HEAT AND COLD IN THE CILIATE INFUSORIA.

REACTIONS TO HEAT AND COLD IN THE CILIATE INFUSORIA.

To explain the movements of organisms toward or from a source of stimulus, we find given almost universally in one shape or another a certain general formula. This is the schema set forth, with unessential variations, by Verworn (1899, pp. 500-502) for the orientation of a ciliate or flagellate infusorian to a one-sided stimulus, and by Loeb (1897, pp. 439-442) for the tropisms of organisms in general. Essentially, the schema is as follows: An agent acting upon the organism from one side causes the locomotor organs of that side to contract either more strongly or less strongly than those of the opposite side.

FIG. 1.*

In the former case (Fig. 1) the animal is turned away from the source of stimulus, till it comes into a position in which the motor organs of the two sides are similarly affected. Then progressing straight forward, it of course moves away from the source of stimulus (negative taxis or tropism). If the motor organs on the side most affected are caused to contract less strongly than those on the opposite side (Fig. 2)

---

*FIG. 1.—Diagram of a negative reaction of an organism, according to the tropism schema. The motor organs which act more effectively are shown more heavily drawn. The more pointed end is the anterior. A stimulus is supposed to impinge upon the organism *a* from the direction indicated by arrows; this causes the motor organs directly affected by the stimulus to beat more strongly, as indicated by the darker shade. The result is to turn the anterior end in the direction indicated by curved arrows. The organism thus occupies successively the positions *a*, *b*, *c*, finally coming into the position *d*. Here the motor organs of the two sides are equally affected by the stimulus, hence there is no further cause for a change of position. The usual forward motion of the organism now takes it away from the source of stimulus, as indicated by the straight arrow at *d*.

7

the organism is necessarily turned with anterior end toward the source of stimulus; then its usual forward movements take it toward the source of stimulus (positive taxis or tropism). Loeb lays especial stress on the *direction* from which the stimulus comes, as it is this that determines which side shall be most strongly affected by the stimulus; otherwise the theory as he sets it forth is essentially like that held by Verworn. Both these authors apply this schema to the movements of organisms to and from many sorts of stimuli, making it a general formula for *taxis* or *tropisms*. Verworn says (1899, p. 503):

Thus the phenomena of positive and negative chemotaxis, thermotaxis, phototaxis and galvanotaxis, which are so highly interesting and important in all organic life, follow with mechanical necessity as the simple results of differences in biotonus, which are produced by the action of stimuli at two different poles of the free living cell.

In the present series of papers the writer proposes to examine the behavior of a number of lower organisms, in order to determine

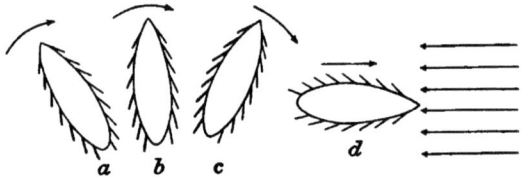

Fig. 2.*

whether the reactions to the usual stimuli take place in accordance with this tropism schema or not, and if not, to determine the real nature of the reaction method. In this first paper we shall deal with reactions to heat and cold.

In his recent series of papers on the reactions of infusoria to heat and cold, Mendelssohn (1902, *a*, *b*, *c*) develops a theory of thermotaxis in accordance with the general theory of tropisms, above set forth. In an earlier paper (Jennings, 1899) the present author, on the other hand,

---

*Fig. 2.—Diagram of a positive reaction, according to the tropism schema. A stimulus coming from the direction indicated by the arrows to the right acts upon the organism *a*. The effect of the stimulus is to cause the motor organs directly affected by it to contract less strongly, as indicated by the lighter shade on the right side of *a*. As a result the animal is turned as shown by the curved arrows, occupying successively the positions *a*, *b*, *c*, *d*. At *d* the stimulus affects the two sides alike, hence there is no cause for further turning, and the usual forward movement of the organism takes it toward the source of stimulus.

gave a brief account of the reactions of Paramecium to heat and cold, according to which these reactions are quite inconsistent with the tropism schema. As the matter is one of considerable interest, and the conclusions reached by Mendelssohn and myself seem quite irreconcilable, I have examined anew the phenomena in a considerable number of infusoria, including Paramecium.

The general phenomena to be explained are well seen in the following experiment, taken from Mendelssohn (Fig. 3). An ebonite trough 10 cm. in length and 2 cm. wide is filled with water containing Paramecia (*a*). Now, by proper methods, one end of the trough is slowly heated to 38°, while the other is kept at the temperature 26°. The

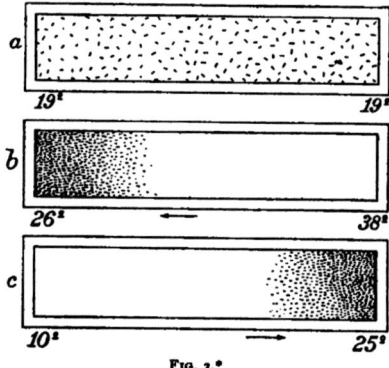

FIG. 3.*

Paramecia soon leave the heated region, traveling away from it in a rather compact mass, and in 5 to 15 minutes they have reached the opposite end (*b*). If now the temperature at the two ends is reversed, the Paramecia travel back to the end from which they came. If the temperature is lowered to 10° at one end, instead of raised, similar results are obtained; the Paramecia leave the cold region, as before they

---

* FIG. 3.—General phenomena of thermotaxis in Paramecium, after Mendelssohn (1902, *a*). At *a* the Paramecia are placed in an ebonite trough, both ends of which have a temperature of 19°. The Paramecia are equally scattered. At *b*, the temperature of one end is raised to 38°, while at the other it is only 26°. The Paramecia collect at the end having the lower temperature ("negative thermotaxis"). At *c*, one end has a temperature of 25°, while the other is lowered to 10°. The Paramecia now gather at the end having the higher temperature ("positive thermotaxis").

left the heated region (c). If one end is heated, while the other is cooled, the Paramecia gather in the intermediate region.

How are these movements to be explained? Mendelssohn applies to the phenomena Verworn's schema for the orientation of a ciliate organism to a one-sided stimulus (see Figs. 1 and 2). As we wish to deal thoroughly with this schema, it will be well to set it forth here, as applied by Mendelssohn to heat and cold, with some fullness.

The temperature being higher at one end of the trough than at the other, that side or end of the animal directed to the heated end of the trough has a higher temperature than has the opposite side or end (see Fig. 4). This difference in temperature causes a difference in the beat of the cilia. In negative thermotaxis the higher temperature causes the cilia to contract more strongly, as indicated by the heavier shade (on the left side) in the figure; hence the animal is turned toward the opposite side, or away from the source of heat, until it comes into a position where the heat acts equally on the two sides. The Paramecium then of course has its anterior end directed from the heated region, and its ordinary swimming carries it away. In positive thermotaxis, on the other hand, the lower temperature causes stronger contractions; hence the cilia on the side next the cold region contract more strongly, turning the anterior end in the opposite direction. The Paramecium then swims away, as a result of its normal forward movement.

FIG. 4.*

Mendelssohn studied the subject primarily from a quantitative standpoint, determining the optimum temperature, the rate of reaction, the effects of different temperatures, etc. For this purpose he constructed a very ingenious and delicate apparatus, which permitted accurate quantitative results. Relying then upon his valuable papers for these matters, I have devoted myself entirely to a study of the mechanism of the reactions. For this purpose an apparatus was used that is similar

* FIG. 4.—Diagram of the thermotactic reaction of Paramecium as conceived by Mendelssohn, after Mendelssohn (1902, b). The heavier cilia on the left side show those contracting most strongly and hence those most effective in turning the organism or driving it forward. In negative thermotaxis the left end would have the higher temperature, causing the cilia of the left side of the organism *a* to beat more strongly. As a result, the organism turns, occupying successively the positions *a, b, c, d*. In the latter position there is no further cause for turning, and the animal swims directly away from the heated end. The same diagram illustrates also positive thermotaxis, if the left end is supposed to be cooled below the optimum.

in principle to that of Mendelssohn, but more easily constructed and permitting exact observation of the organisms with the microscope, though otherwise much less elegant than Mendelssohn's. This apparatus is shown in Fig. 5. It consists essentially of three glass tubes, of 8 millimeters bore, which are supported in a horizontal position, side by side, by passing them through auger holes in a block of wood. The tubes are one inch apart and are placed exactly at the same level, so that a glass slide rests equally on all three. To the two ends of each of these rubber tubes are attached. The rubber tubes from one end pass upward into vessels of water raised on a shelf above the level of the apparatus. From the other end the rubber tubes

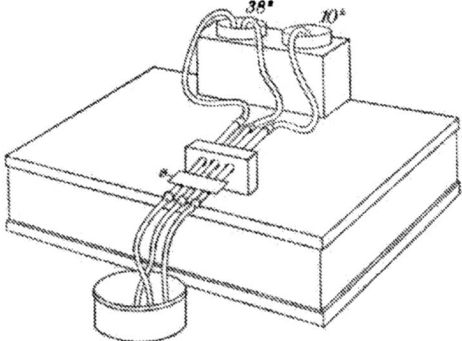

FIG. 5.*

pass downward into a waste pail, thus acting as overflow tubes. A trough, or slide (*s*), containing infusoria, is placed on the three glass tubes; the water in the vessels on the shelf is heated or cooled to any desired temperature, and is then siphoned off and allowed to flow downward through the glass tubes. The rate of flow is controlled by pinchcocks. In this manner heated water can be caused to flow beneath one end of the slide, cold water beneath the other. The slide being thus unequally warmed, the reactions of the organisms can be observed. The rubber tubes leading from the hot and cold vessels can be interchanged, so that the temperature at either end or the middle of the slide can be at once changed and made high or low, without the

---

* FIG. 5.—Apparatus used for testing reaction to heat and cold. For description, see text.

slightest disturbance to the slide or trough containing the organisms. The plan of this apparatus is taken from that of Mendelssohn. It can be readily constructed in an hour or less, and gives essentially the same results as Mendelssohn's more elaborate arrangement. With the use of specially constructed thermometers, such as were employed by Mendelssohn, exactly the same quantitative work could be done. The present apparatus has the advantage that it is possible to place a mirror beneath the glass slide or trough bearing the organisms, and thus to observe the movements of the latter with the microscope by the aid of reflected light. With the long-armed Braus-Drüner stand the whole extent of the trough can be examined at ease, and the movements of the organisms accurately observed with the stereoscopic binocular.

As a trough I usually employed a glass slide, to which strips of glass 2 mm. in diameter had been cemented, making a trough 3 inches long, about two-thirds of an inch wide, and 2 mm. deep. In some of the experiments the trough was covered with a glass plate; in others it was left open. Both methods have their advantages and disadvantages.

To realize the exact conditions under which the organisms are acting it is necessary to consider a further question: What is the precise nature of the stimulating agent in these experiments? Are we dealing with radiant heat or with conducted heat? If we are dealing primarily with radiant heat, of course currents in the water have no effect on the distribution of the stimulating agent. If, on the other hand, we are dealing with conducted heat, if the stimulating agent is the heated or cooled water, then the conditions are different. Local currents will cause local variations in the distribution of the heated water. It is evident, I think, that the second alternative is in all probability the correct one. Certainly in a bath-tub or in a long vessel of any sort in which the water is heated at one end and not at the other, it is possible by producing currents to vary the distribution of the heated water and to perceive with the hand that it is this heated water which acts as the stimulus.

The importance of these considerations is evident when we take into account the fact that the ciliate infusoria are always accompanied by currents of typical character, having a definite relation to the form and orientation of the animal's body. As a result of these currents, the infusorian becomes not a mere passive recipient of stimulations, but an active agent, determining by its activity how and in what part of the body it shall be affected by stimuli. This may be illustrated by a diagram (Fig. 6) showing the typical currents produced by the cilia of Paramecium and the effect produced by these currents upon the distribution of the heated (or cooled) water. The temperature is conceived to be greatest to the right of the figure and to fall off regularly toward the left, the lines indicating regions of equal temperature. The last line to the left is marked 28°, this being about the threshold temperature for the negative reaction of Paramecium, according to Mendelssohn. The space about the Paramecium (without lines) is at a temperature below 28°—say at the room temperature—so that it does not act as a stimulus to cause movement. Now, as the diagram

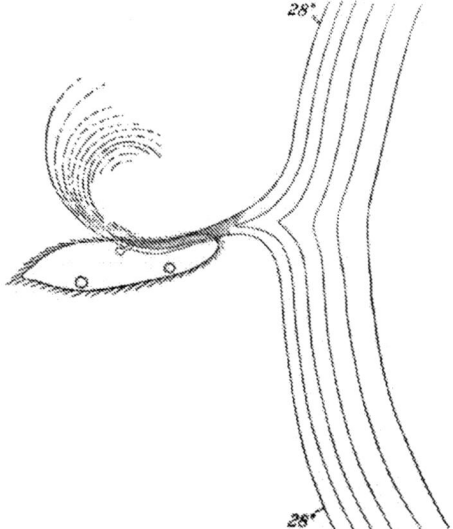

FIG. 6.*

shows, a cone of water is drawn toward the anterior end of Paramecium, from a considerable distance away, necessarily therefore including water above the threshold temperature of 28°. This cone or

---

*FIG. 6.—Diagram of currents produced by the cilia in Paramecium when the animal is nearly or quite at rest. At right of the line marked 28° the temperature is above the optimum (above 28°) while at the left of this line it is at the normal or optimum temperature. The heated water first reaches the Paramecium at the anterior end on the oral side, passing down the oral groove to the mouth.

vortex of water passes as a slender stream along the oral groove of Paramecium to the mouth. Consequently, water heated above the threshold temperature reaches the Paramecium in this region before it touches the body elsewhere. The result is thus a stimulation on the oral side of the body, not elsewhere.

Thus the way in which the organism is stimulated depends not exclusively on the physical laws of the distribution of heat, but upon the activity of the organism; and the method of reaction, as we shall see, is of a corresponding character.

It is not difficult to observe the distribution of the currents above described if one adds to the water on one side of the nearly or quite quiet infusorian a cloud of very finely ground India ink. The same results are obtained with other infusoria; in Stentor, in Bursaria, and in some of the larger Hypotricha the results are particularly striking. Of course if the India ink, or the surface of threshold temperature, is advancing obliquely to the axis of the infusoria, the results are more complicated, and a diagram such as we have in Fig. 6 is not easy to construct. But the result is uniformly to bring the stimulating agent to the peristome before it reaches any other part of the body. It is not possible to observe directly the distribution of water of different temperatures, but under the influence of currents this of course follows, essentially, the same laws as do fine particles suspended in the water.

Another factor which it is important to take into consideration in studying the effects of heat or other agents on the infusoria is the greater sensitiveness of the anterior end and oral surface (or peristome) as compared with the remainder of the body. This the present writer has demonstrated for the anterior end by direct mechanical stimulation in a considerable number of infusoria (Jennings, 1900), while Roesle (1902) has shown a similar high comparative sensitivity for the peristome region. The difference is such that in many cases where the animal is completely enveloped by a stimulating agent (as by a chemical, or by warm or cold water) there is reason to think that the reaction given is due to the stimulation at these regions alone. In other words, the stimulus reaches its threshold value for the anterior end and the region about the mouth much before it reaches this value for the rest of the body. This consideration has an important bearing on the theory which is frequently maintained, that the directive action of a stimulus is due to the difference in its intensity on the two ends or sides of the organism. Even if a stimulating agent acts, *per se*, slightly more strongly on the posterior end than on the anterior end of an infusorian, there is reason to think that the reaction would be conditioned entirely by the stimulus at the anterior end, this reaching its threshold value before the stimulus elsewhere produces any effect. Corresponding statements could be made with relation to the oral and aboral sides. Of course, owing to the course of the currents above described, any stimulating agent whose distribution is affected by currents in the water will usually reach the anterior end and oral side first in any case.

Summing up, we find (1) that the threshold intensity of a stimulating agent whose distribution is affected by currents in the water will reach the anterior end and oral side of the organism before it reaches other parts of the body; (2) that the anterior end and oral surface are more sensitive than the rest of the body, so that the threshold value for stimuli is less here than elsewhere.

We may now proceed to an account of the observed method by which some of the organisms react to heat and cold.

Oxytricha fallax: This is one of the most favorable of the Ciliata for determining the method of reactions to stimuli, for two reasons. (1) It is easily procurable in large numbers, occurring in cultures of the same sort that produce Paramecium, and in equal abundance. (2) It does not, as a rule, revolve rapidly on its long axis, as Paramecium does, but usually creeps with its oral or ventral side against a surface, so that it is not difficult to observe the relation of the reaction movements to the differences in the sides of the body.

When water containing a large number of Oxytrichas is placed in the trough and one end of the trough is heated by passing warm water through the tube which underlies it, the Oxytrichas gradually pass toward the opposite end of the trough, forming a dense assemblage with a rather sharply defined edge toward the heated side. If the end at first heated is now cooled and the opposite end heated, the organisms pass back to the end from which they first came. Similar results are obtained by making one end very cold; the animals gather in an optimum region, avoiding both too great heat and too great cold.* The phenomena are identical with what is to be observed in the case of Paramecium, save that it requires somewhat longer for the Oxytrichas to move from one end of the trough to the other, and the progress in a definite direction is not so steady as we find it in Paramecium.

If the movements of the individuals are observed we find them to be as follows: Near that end of the trough where the temperature is

---

* Many quantitative data for various infusoria are given in the valuable papers of Mendelssohn. As the object of the present paper was not to obtain quantitative data, but to determine just how the animals acted, absolute temperatures are not recorded. In every case the experiments were so varied as to use at times temperatures to which a reaction was hardly noticeable; at other times more extreme temperatures, up to those which were destructive.

raised above the threshold value the animals begin to move about rapidly. At first view this movement seems to be quite irregular, as Mendelssohn describes it in Paramecium. But exact observation of

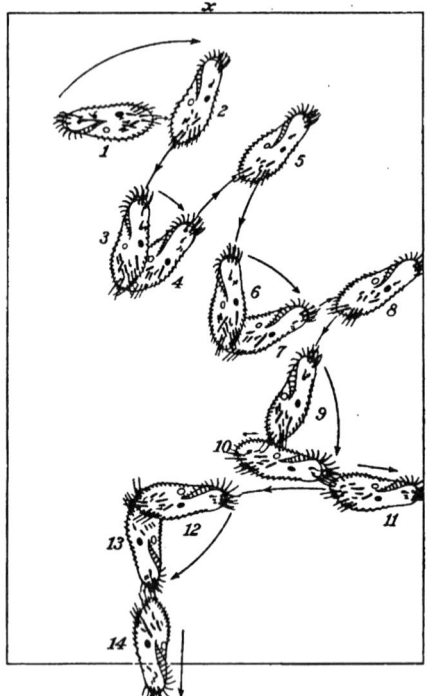

FIG. 7.*

the individuals taken separately shows that this movement is not so entirely irregular as it at first appears. Most of the animals swim

---

* FIG. 7.—Method by which *Oxytricha fallax* reacts to heat or cold. The figure represents one end of a trough or slide, which is heated from the end *x*. An Oxytricha in the position 1 is reached by the heat coming from the end *x*. The

backward, circling at the same time toward the right or aboral side, as shown in Fig. 7. This lasts but a moment; then the animal swims forward, at the same time turning to the right or aboral side. That is, the individuals give the typical motor reaction, as described in the fifth of my studies (Jennings, 1900). This reaction is repeated many times, as long indeed as the animal remains in the heated region. But of course this movement scatters the animals rapidly. Those that strike against the end or sides of the trough repeat the reaction above described, backing, turning to the right, then going forward (Fig. 7 at 8, 9, 10, 11). They thus become directed in some other way. Those that are directed away from the heated region pass into cooler water and hence no longer give the reaction, but continue their course (Fig. 7 at 13, 14). The result is that the individuals which swim *away* from the heated end continue their course, while those starting in any other direction are stopped and turned (through the motor reaction), until they too get started away from the heated region. Thus after a time there is a steady stream of organisms swimming or creeping away from the heated end, while there is no regular movement in any other direction. In this manner arises the orientation of the animals, with anterior ends directed away from the heated region.

The movements of the individuals are exactly as above described even when the heat is applied some distance from the region where the animal is found and gradually approaches it from one side. The animal by no means turns directly away from the heated region, but repeatedly gives the backing and turning reaction till it is finally moving in a direction which takes it out of the heated region.

How is this continued backing and turning to be accounted for on the theory of direct action on the locomotor organs of the two sides as maintained by Mendelssohn? This author speaks in the case of Paramecium merely of "disordered" movements when the reaction first

---

animal reacts by turning to the right and backing (1, 2, 3), turning again (3-4), swimming forward (4-5), backing (5-6), turning again to the right (6-7), etc., till it comes against the wall of the trough (8). It then reacts as before, by backing (8-9), turning to the right (9-10). This type of reaction continues as long as the Oxytricha is in the heated region, or as long as its movements carry it either against the wall or into the heated region. When it finally becomes directed away from the heated region (13), as it must in time if it continues its reactions, it swims forward, and since it is no longer stimulated, it no longer reacts. When large numbers of animals react in this way, in the course of time nearly all become pointed in the same direction, as at 13 or 14, so that a marked "orientation" is produced. Thus orientation is produced by "exclusion," due to the fact that the organism is prevented, either by the heat or the walls of the trough, from swimming in any other direction.

begins, and thinks this is due to "individual differences, or to ill-defined internal causes, or perhaps rather to the heterogeneity of the medium in which they find themselves" (Mendelssohn, 1902, *c*, p. 492), and that it has nothing to do with thermotaxis proper. This is typical of many of the statements made concerning the behavior of the lower organisms; the movements, so long as they do not agree with the preconceived schema, are cast aside as disordered, and attention is called only to the movements that do not conflict with the theory. Thus Mendelssohn says that this disordered movement "ceases immediately as soon as the thermotactic action manifests itself" (*l. c.*, p. 492). This is true merely because the thermotactic action is conceived to begin only after the organism has, through the movements above described, gotten itself into such a position that it moves away from the heated region. Of course if all movements except those after orientation has occurred are thrown out of consideration, the orientation can be accounted for in any way desired.

In Paramecium, for which alone Mendelssohn attempts to give an account, based on observation, of the mechanism of the thermotactic response, the exact character of the movements is undoubtedly difficult to observe. This animal is nearly cylindrical in section; the oral side is very slightly marked, the movements are rapid, and the animal continually revolves rapidly on its long axis, so that observation of the relation of the direction of turning to the differentiations of the body is very difficult. In Oxytricha and other Hypotricha these difficulties are almost absent; the body is markedly differentiated; the movements are less rapid, and, most important of all, there is usually no revolution on the long axis. It is unfortunate therefore that Mendelssohn included none of the Hypotricha among the organisms which he studied. With careful observation of the movements of individuals the mechanism of the reactions is in these animals absolutely clear.

A crucial test of the theory of direct orientation as maintained by Mendelssohn is given by observation of the direction in which the animals turn in becoming oriented. Mendelssohn (1902, *c*, p. 492) says that after the disordered movements "the movements executed to place the body in orientation are rather movements of rotation." This could hardly be otherwise, but the important question for deciding as to the nature of the reaction is, How does the rotation take place? Is it determined by the direction from which the heat comes, as required by Mendelssohn's theory, or is it determined by the differentiations of the animal's body? This point is a decisive one for interpreting the nature of the reaction. Suppose we have an Oxytricha in the position *a-a*, Fig. 8, and heat is applied in such a way as to reach the organism from the direction indicated by the straight arrows. The heat is supra-optimal, so that the organism moves away from it. In what direction will the organism turn in order to reach the position of orientation $b\text{-}b$? According to the theory of Mendelssohn, that the orientation is due to an increase of the effective beat of the cilia on the side from which the heat comes, the animal must turn in the direction indicated by the arrow x, and this is of course what one would naturally expect, since this is the most direct method of becoming oriented. But as a matter of fact the organism turns in the opposite direction, as indicated by the arrow y, thus demonstrating the incorrectness of the theory that orientation is due to increase of the effective beat of the cilia on the side from which the heat comes. I have made this observation hundreds of times, not only upon Oxytricha, but on other Hypotricha and on infusoria belonging to other groups (see below). The direction of turning is determined, under the heat stimulus, by the differentiation of the animal's body. Oxytricha turns to the right, without regard to the direction from which the heat comes. This is very striking when the trough is covered and part of the animals are creeping on the cover-glass with ventral side up, while the remainder are creeping on the bottom of the trough with ventral side down. When stimulated by heat approaching from one side, all the members of the first group will be observed to turn counter clock-wise, while those of the second group turn in the same direction as the clock hands; that is, each specimen turns toward its right side.

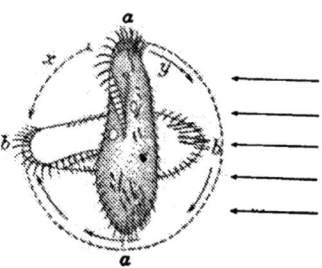

FIG. 8.—Method of orientation in Oxytricha. For details, see text.

For becoming completely oriented an animal in the position $a\text{-}a$ in Fig. 8 usually requires a number of reactions, as indicated in Fig. 7, but the turning in every case is as indicated by the arrow y (Fig. 8).

After it has become oriented with the anterior end away from the source of heat, Oxytricha by no means maintains this position with rigidity; on the contrary the individuals shoot back and forth, in a way that might be anticipated from the method in which the reaction occurs. They thus form groups here and there, which gradually move away from the heat, much as is described by Mendelssohn for *Paramecium bursaria*. With a large number of individuals a general orientation is evident, however, after the experiment has been some time in progress.

If ice water is used as the stimulating agent in place of heated water, the phenomena to be observed are practically identical with those above described. The organisms leave the colder region, giving the same reaction as with heated water. As the cold has the additional effect of decreasing the movements, many individuals are immobilized by a very low temperature before they have succeeded in escaping from it, so that they remain in the cold region. The reaction is thus less clearly defined than that to heat.

Oxytricha æruginosa: This organism, though smaller, is in some respects more favorable than *O. fallax* for observing the method of reaction. This is because the individuals are more inclined to swim freely through the water, so that their progress away from or toward the heated region is more rapid than in *O. fallax*. *O. æruginosa*, further, even when moving freely through the water, either does not revolve on the long axis at all or revolves only very slowly. In consequence of this it is easy to determine the relation of the direction of turning to the differentiations of the body.

The reaction to heat and cold is in essentials identical with that of *O. fallax*, and this reaction is repeated till the animals are carried into a region where the temperature is not such as to cause the reaction. Those that *are* carried into such a region will of course be swimming away from the stimulating region; hence, in a large number of individuals there is an evident orientation, with anterior end directed away from the source of heat or cold. All the conditions and details as to the production of this orientation are as set forth above for *O. fallax*.

Stylonychia mytilus: This large Hypotrichan is still more favorable for the study of the movements of individuals under the stimulus of heat or cold coming from one side than are the two species of Oxytricha. But I have found it less easy to obtain in large numbers, and for this reason have not chosen it for the detailed description of the reaction. Where comparatively few specimens are available, the movements of individuals are easily studied, but there is little impression of any real orientation, such as one gets clearly when large numbers are used.

The movements of the individuals are like those described for *Oxytricha fallax*. The animal in reacting always turns to its right, without regard to the relation of this to the direction from which the heat or cold is coming. With an organism of the large size of

Stylonychia this is very evident. This turning to the right under the stimulus of heat and cold in Stylonychia has already been described by Pütter (1900), incidentally to his study of the effect of contact stimuli in this organism.

Stentor cæruleus: Mendelssohn includes in his paper a note stating that positive and negative thermotaxis occur in some species of Stentor, and giving the optimum; but he made no study of the mechanism of the reactions in this animal. Had he done so, it seems to me that he could not have maintained his theory of the way in which the reaction takes place.

When one end of the trough is warmed the Stentors near that end begin after a few seconds to move about more rapidly. In most cases the movement is as follows: The animals swim backward some distance, then turn toward the right aboral side and swim forward (the typical motor reaction). Thus the general effect is as of an irregular movement in all directions. Those individuals which swim forward toward the other end of the slide pass out of the heated region; hence the motor reaction no longer takes place, and the animals continue to swim forward. Those which start in any other direction do not escape from the heated region, and therefore soon give again the motor reaction, backing and turning again to the right. Thus only those that swim away from the heated region contique their course; the others are stopped and turned until finally they too get started in the same direction. Therefore, after a period of apparently disordered swimming, there is an evident orientation of many individuals, with anterior ends away from the heated region. This orientation is caused as it were by exclusion; in animals swimming in any direction but one the motor reaction is produced, so that only this direction can be maintained. After a time, therefore, a large proportion of the individuals are swimming in this direction, with a common orientation.

Thus the direction in which the animals turn is determined, as in the Hypotricha, by the structure of the body, and not by the direction from which the heat comes.

Those outside the region where the heat has reached the threshold temperature often swim for some distance toward the heated region; then arriving at a point where the heat is effective, they give the motor reaction, backing and turning to the right. They are thus prevented from entering the heated region.

If the temperature is rapidly raised, the animals may not succeed in escaping from the heated region until they are injured. In this case the specimen contracts strongly and swims backward a long time. It becomes distorted, places the disk against the bottom or other surface, becomes motionless, and finally dies.

Fixed specimens react less readily to heat than do free-swimming specimens. They do not orient themselves with reference to the direction from which the rise in temperature comes. They may remain extended normally, carrying on the usual activities, after the temperature has risen beyond the point which sets the free specimens in rapid reaction. But as the temperature rises they repeatedly bend over into a new position (bending toward the right aboral side), then contract strongly, and finally free themselves from their attachment. Thereupon they behave like other free individuals.

Spirostomum ambiguum: In this large ciliate the reactions to heat and cold take place in essentially the same manner as is described above for Stentor and the Hypotricha, so that it is not necessary to describe the phenomena in detail. The organism reacts to heat or cold by backing and turning toward its aboral side; and this whether the change in temperature is uniform over the entire surface of the animal or whether it approaches from one side. The movements of the animal are slow, and under the Braus-Drüner stereoscopic microscope its method of reaction is very clear. There is little marked common orientation at any time, however; this being due to the slowness of the movements and the frequency of repetitions of the motor reaction.

Bursaria truncatella: In this very large infusorian, in which certain differentiations of the body are visible even to the naked eye, the method of reaction to heat and cold is observed with the greatest ease. But orientation of a large number of individuals in a common direction is hardly to be noticed, though if Bursaria could be obtained in such numbers as Paramecium or Oxytricha, perhaps an indication of orientation would be noticeable in spite of the slowness of movement.

Bursaria is very inactive, often remaining quiet for long periods. It swims slowly, and frequently creeps along the bottom with ventral side down, but may also swim freely through the water, revolving to the left. If the temperature of the trough is raised at one end, the animals in this region that are moving freely through the water swim backward, turn to the right, and swim forward. This may be repeated till the organism passes out of the heated region. Rather more frequently, however, the animal, after thus reacting once or twice, sinks to the bottom and places its ventral side against the surface. It now conducts itself in the same manner as do the other individuals in this situation, as will be described.

The individuals which are resting against the bottom (usually the majority of those in the trough) react as follows: They begin to swim backward, keeping the ventral side down and at the same time circling toward their own right sides. They thus describe rather narrow circles.

This continues until the heat becomes destructive—the animals cease circling, become quiet, and finally disintegrate. The reaction of those individuals which are resting or creeping on the bottom is thus not of a character to save them from destruction.

Specimens which are by chance moving along the bottom from a cool region toward the warm region do not escape; they merely stop and begin to circle backward to the right when they reach the heated spot, and continue this till they die.

Thus the reaction of Bursaria to heat, while of the same general character as that of other infusoria, must be accounted very imperfect, since it hardly results in orientation at all, and does not preserve the animals from destruction.

Paramecium caudatum: [*] In the second of my studies (Jennings, 1899, pp. 334-336) I gave a brief account of the way in which, according to my observations, Paramecium reacts to heat and cold. From my more recent studies I can confirm this account. But as Mendelssohn has recently come to different conclusions for the temperature reaction of this animal, and as he misunderstands certain points in my brief description, it seems desirable that I should supplement the account previously given in order to make it clear.

Paramecium reacts to heat and cold in essentially the same manner as is described above in detail for Oxytricha. When the higher or the lower temperature advances from one side the animals swim backward, turn toward the aboral side, and swim forward again. They continue this until the movement brings them into a region of more moderate temperature. Paramecium reacts more readily than Oxytricha, the reactions are repeated at shorter intervals, and the movements are more rapid, so that a common orientation of many individuals swimming away from the region of higher or lower temperature is more quickly produced and is more striking to the eye. It results farther from this more rapid movement, as well as from certain other factors, that the method of reaction in Paramecium is much less easily observed than in any of the other infusoria described. Indeed, Paramecium is one of the most unfavorable forms obtainable for a study of reaction methods, and it is, I believe, due largely to the fact that this animal is usually employed for such study that progress has

---

[*] The common Paramecium, which appears everywhere in immense numbers in decaying vegetation, receives from different authors sometimes the name *Paramecium aurelia*, used by Mendelssohn; sometimes the name given above. I use the name *caudatum* because it appears to me to be the correct one, but there is no reason for considering the animals thus differently denominated to be really different.

been so slow in appreciating the real nature of the reactions of the infusoria. If Stylonychia or Oxytricha or any other of the Hypotricha had been taken as the usual type for study on reactions, many of the theories now maintained could never have been put forth. The body of Paramecium is comparatively little differentiated, so that it is difficult to distinguish oral and aboral sides, and, to multiply this difficulty many times, the animal revolves rapidly on its long axis, so that oral and aboral sides never retain for two successive instants the same position. It is not wonderful, therefore, that the method of reaction by turning toward the aboral side was not observed in the first investigations on Paramecium and that many still find it difficult to observe. Nevertheless, it was on Paramecium itself that this reaction method was first observed (Jennings, 1899), and its existence was confirmed later on the organisms where its observation presents no difficulties. Aside from the direct observations of the method of reaction, the following facts throw light on the way in which the collections take place.

FIG. 9.*

As described in the second of my studies (Jennings, 1899, pp. 314, 315), the collecting of Paramecia in regions of optimum temperature may be produced in the following manner: The infusoria are mounted in water which is above the optimum temperature (say 30°) on a slide beneath a cover glass supported at its ends by glass rods. Into this slide is introduced with the capillary pipette a little cooler water (say at 24°), which covers a small circular area in the center of the slide. Very soon the Paramecia have collected in this region till a dense group is formed. The same result may be obtained by placing a drop of ice water on the top of the cover glass of a slide of Paramecia which has been warmed considerably above the optimum temperature. (Fig. 9.)

Are these collections due to the orienting of Paramecium by the heat, as maintained by Mendelssohn for thermotaxis in general? Observation shows that they are not; that on the contrary the Paramecia gather in the optimum region in the same manner as they gather in a drop of weak acid, as described in my studies. The Paramecia on the heated slide are swimming rapidly in all directions. They do not change their course or become oriented in the least when a spot in a certain part of the slide is cooled. But as a consequence of their

*FIG. 9.—Collection of Paramecia due to the reaction to temperature change. The slide rests on a vessel of water at a temperature of 45°. An elongated drop of ice water is placed on the upper surface of the cover glass. The Paramecia quickly collect beneath the drop of ice water.

rapid movements many of them by chance enter the cooler region. They do not react at all as they enter, but continue across. On coming to the other side of the drop, however, they *do* react, by backing and turning toward one side (the aboral). They react whenever they come to the boundary of the cooled region; hence they do not leave it. In every respect their behavior is like that seen when Paramecia collect in a drop of weak acid, and I believe there is no longer anyone who holds to the orientation theory for the gathering of Paramecium in chemicals.

As in the case of chemicals, it may be demonstrated to the eye in the following manner that the method above described suffices to account for the gatherings. On the *upper* surface of the cover glass is marked a small ring in ink. By confining the attention to this ring it is easily seen that in the heated preparation of Paramecia many individuals cross the ring every instant, so that, if these could all be stopped *in* the ring, a dense aggregation would soon result. Then the region within the ring is cooled by placing a drop of ice water on the cover above it. The Paramecia continue to swim just as before, save that they no longer pass *out* of the ring after swimming in, as they did at first. In this way a dense collection is soon formed.

Mendelssohn (1902, *b*, p. 487) finds it inexplicable why the Paramecia should form dense aggregations at the optimum temperature. He says that they execute "only some insignificant movements" in this region, not swimming away. On the theory of thermotaxis held by Mendelssohn this is perhaps inexplicable, but this, it seems to me, is only because the theory is incorrect. Such collections are due to precisely the same factors as the rest of the reaction to heat and cold and are clearly intelligible when the nature of the reaction, as described above, is taken into consideration.*

In a former paper (Jennings, 1899, p. 336), after giving a brief account of the reaction method above described, I pointed out that this method does not demand a sensitiveness to such minute differences in temperature as does Mendelssohn's theory, and that therefore the sensitiveness to temperature differences may have been overestimated.

---

* Mendelssohn (1902, *b*, p. 487) supposes that I would explain these gatherings at the optimum temperature through the collection of Paramecia in CO_2 produced by themselves, and shows that this would not account for the phenomena observed in these cases, though he confirms the fact of the collections in CO_2. But I have by no means maintained that such collections can be produced only by CO_2; on the contrary, I have given an account of many different agencies that will give rise to such collections, and have especially described the fact that collections are formed in a warmed region through exactly the same reaction by which they are formed in CO_2. (Jennings, 1899, p. 315.)

Mendelssohn (1902, *a*, p. 406) misunderstands my ground for this statement. He supposes that I hold that the Paramecia do not react to differences in temperature less than that existing in a certain illustrative experiment, where one end of the slide was resting on ice, while the other was heated to 40°. This experiment was purely for the purpose of bringing the phenomena of thermotaxis concretely before the attention of the reader; its details had no special significance. I have not the slightest reason for doubting the entire accuracy of the quantitative experimental results set forth by Mendelssohn, and consider them a most valuable addition to our stock of exact data. But the calculation of the sensitiveness of the organisms concerned, from these experimental results, involves a certain interpretation as to the reaction method, and it was this interpretation that I called in question. Mendelssohn, in accordance with his general theory, holds that the reaction is due to the difference in temperature between the two ends of the organism, and he calculates that this difference in temperature could amount, in the case of Paramecia, to but 0.01° C. According to the reaction method which I have described above, however, it is not the difference in temperature between the two ends of the same individual that causes the reaction. Consider a slide cooled below the optimum at the end *a*; above the optimum at the end *b* (Fig. 10), the optimum temperature for the Paramecia being between the lines *x* and *y*. The animal may swim a considerable distance from a position *y*, at one side of the optimum, to a position *x*, at the other side of the optimum, before it reacts (by backing and turning, etc.) at all. We have no ground for maintaining then that it perceives any less differences in temperature than that between the lines *x* and *y*, and this difference will be much greater than that between the two ends of the animal. A similar diagram could be made for the case where the temperature is raised or lowered only at one end of the slide. It seems to me correct, therefore, that the sensitiveness to temperature differences has probably been much overestimated. The only way that it could be estimated would be by observation of individuals to determine the extent of the stretch *x-y* over which they pass before reacting, and to calculate the difference in temperature between the ends of this stretch. It would of course be very difficult to do this with accuracy.

Mendelssohn's view that it is the difference in temperature between the two ends of the same individual that determines the reaction is not

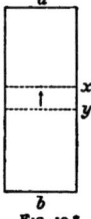
Fig. 10.*

---

* Fig. 10.—Diagram illustrating conditions necessary for determining the sensitiveness of Paramecia to differences in temperature. See text.

only rendered inadmissible by the reaction method above described, but it is rendered *a priori* improbable by certain other considerations. First we have the fact that the anterior end is much more sensitive than the posterior. Of course it is impossible to measure this difference in sensitiveness, yet the experiments with mechanical and chemical stimuli show that it is great. In many infusoria, while the slightest touch at the anterior end causes a pronounced reaction, it requires a strong stroke at the posterior end to produce even a slight reaction. (See Jennings, 1900, pp. 238, 243, 251.) Owing to the much greater sensitiveness of the anterior end, it is probable that, with the posterior end but 0.01° warmer than the anterior, the reaction, if any, would be due to the temperature of the anterior end. In other words, there is reason to suppose that the threshold temperature for the anterior end would be considerably lower than that for the posterior end. If this is true the usual temperature reactions would be throughout due primarily to stimulation at the anterior end; and the reaction, as we have seen, is of just the character which would be expected from this. The first stage in the reaction is to *swim backward*, and this is true also when the animal is dropped directly into water of uniformly high or low temperature, so that the temperature of the anterior end is no greater than that of the posterior end. There is no explanation for the swimming backward under these circumstances on the theory that accounts for thermotaxis by the different temperature of the two ends.

A second factor which must be taken into consideration relates to the currents produced by the cilia of the organism itself. As shown above (p. 13), the water of a higher temperature (supposing that we are dealing with the reaction to heat), would as a rule first reach the anterior end and pass at once down the oral groove, on the oral side (Fig. 6). The natural result therefore would be a turning toward the opposite or aboral side, and this is exactly what we find takes place. We should therefore not expect the organism to turn *directly* away from that end of the trough from which the heat comes, for the heated water may not reach the Paramecium from that side at all.

As will be seen, the facts adduced in the last paragraph are not inconsistent with the idea that the organism turns directly away from the side stimulated. It is the oral side which is, as a rule, stimulated, and the organism turns toward the aboral side. We seem thus to obtain a most gratifying union of two apparently opposed views. But the reactions to certain other stimuli do not admit of such a union. This is notably true of the reactions to mechanical stimuli, as shown in a previous paper (Jennings, 1900), and of the reactions to light, to be described in the following paper.

SUMMARY.

The ciliate infusoria react in the same manner to heat and cold as to most other classes of stimuli; the response on coming into a region where the temperature is above or below the optimum is by backing and turning toward a structurally defined side, followed by a movement forward. This reaction is repeated as long as an effective supraoptimal or suboptimal temperature continues. The result is to prevent the organisms from entering regions of marked supraoptimal or suboptimal temperature, and to cause them to form collections in regions of optimal temperature. The common orientation of a large number of individuals sometimes produced in this way is an indirect result of the method of reaction. Since movement in any other direction than a certain one is stopped, the organisms after many trials come into this direction. Orientation is therefore by "exclusion," or by the method of trial and error. In many of the organisms orientation is not a noticeable feature of the reaction.

SECOND PAPER

—

REACTIONS TO LIGHT IN CILIATES
AND FLAGELLATES.

REACTIONS TO LIGHT IN CILIATES AND FLAGELLATES.

In the reactions to light we are dealing with a stimulating agent which differs in one very important respect from chemicals and from heat or cold. The distribution of the agent with which we are concerned is not affected by the currents of water produced by the organism; hence there is no tendency for one side or part of the organism to be more strongly affected than the rest, as was found to be the case for chemicals and for heat and cold. This peculiarity light shares with the electric current and with radiant heat. The conditions demanded for immediate orientation through direct action of the agent on the locomotor organs, in the manner required by the general theory of tropisms as set forth in the foregoing paper (p. 7), are therefore present. In a recent paper Holt & Lee (1901) have attempted to show that the reactions of organisms to light actually take place in accordance with this theory.

We shall examine the reactions to light in *Stentor cæruleus* and in certain flagellates, in order to determine whether they take place in accordance with the tropism schema, and, if not, just how they do occur and on what factors they depend.

THE CILIATA.

STENTOR CÆRULEUS.

As is well known, very few of the ciliate infusoria react to light. Light reactions have been described by Engelmann (1882, *a*) for several chlorophyllaceous ciliates; by Verworn (1889, Nachschrift) for *Pleuronema chrysalis;* and by Davenport (1897) and Holt & Lee (1901) for *Stentor cæruleus*. In none of the ciliates have the reactions been described in sufficient detail to enable us to determine their exact nature.

In *Stentor cæruleus* the reaction to light manifests itself in the culture dish by the usual aggregation of the organisms at the side away from the window. If a number of Stentors are removed to a watch glass or trough, and this is placed near a window or other source of light, most of the Stentors are soon found on the side of the vessel away from the light. If one-half of the glass is shaded by a screen, most of the Stentors are soon found in the shaded half. *S. cæruleus* thus shows the phenomenon usually called negative phototaxis.

It is to be noted that not *all* the Stentors are to be found on the side away from the light, or in the shaded half of the vessel. On the contrary, a considerable fraction of the whole number will usually be found swimming about in all parts of the dish, or at rest in the lighted portion. The light reaction is thus somewhat inconstant, and varies among different individuals. It varies considerably with Stentors of different cultures; from some cultures almost all the individuals show it, while from others it is barely noticeable. This variability and inconstancy run through all manifestations of the light reaction in Stentor.

A word further needs to be said as to the behavior of individuals which are not free-swimming, but are fixed by the posterior end. Such individuals do not react at all to light. When light is thrown on them they remain in the positions in which they are found at the beginning, neither contracting nor in any way changing their position. No matter whether the light is weak or strong, and without regard to the direction from which it comes, fixed Stentors give no reaction and show no orientation with reference to light. The contact reaction apparently inhibits the light reaction completely. We shall therefore omit the fixed individuals from consideration in the remainder of the account, confining attention to the free-swimming specimens.

The typical motor reaction of Stentor, by which it responds to most stimuli, is as follows: The Stentor stops or swims backward a short distance, then turns toward the right aboral side, and resumes its forward motion. This is the reaction which is produced by strong mechanical stimuli, by heat, and by chemical stimuli, acting upon the anterior end or upon the body as a whole.

How is the reaction to light brought about? To answer this question it is best to arrange experiments in such a way as to distinguish as far as possible the effect due to unequal illumination of different areas from the effect due to the direction from which the light is coming.

In order to produce strong differences in illumination in different areas of the space in which the Stentors are found, a flat-bottomed glass vessel containing many Stentors in a shallow layer of water was placed on the stage of the microscope, in a dark room. From beneath strong light was sent upward through the opening of the diaphragm, by throwing the light from the projection lantern (using the electric arc light) on the substage mirror. By this the light was directed upward through the vessel containing the Stentors. Thus a small, definitely bounded circular area was illuminated, while the rest of the vessel remained in darkness. A black screen was usually placed over the diaphragm opening of the microscope in such a way as to shade one-half of the circular area, making a sharp line (*x-y*, Fig. 11) dividing the light from the darkness. A mirror was placed above the microscope, inclined in such a position as to project the image of the Stentors, very much magnified, on the ordinary vertical screen used for receiving lantern slide views. Thus the behavior of the Stentors could be studied with great ease on the screen.

The heat from the lantern was cut out, so far as possible, by placing between it and the mirror of the microscope a glass cell three inches thick, filled with cold water. In this manner the heat was excluded to such an extent as to fall below the threshold for the stimulation of Stentor by heat. This was demonstrated by comparing the reactions of Stentor with those of Paramecium. Stentor is less sensitive to changes in temperature than is Paramecium; this was clear in my experiments on the reaction to heat. Paramecium does not react at all on passing into the area illuminated by the lantern, but swims about indifferently in both the dark and the light parts of the dish, showing that the heat produced is below the threshold for Paramecium; it must then be below the threshold for Stentor.

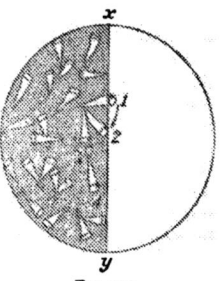

Fig. 11.*

The free Stentors in the unlighted part of the vessel swim about at random. Many individuals thus come by chance to the line *x-y*, Fig. 11, where they would pass into the lighted area. These at once back a little, then turn toward the right aboral side, and swim forward again. The turning toward the right aboral side is usually through an angle sufficient to direct the Stentor away from the lighted area (see 1, 2, 3, 4, Fig. 11); if it is not, the Stentor repeats the reaction until, after one or two trials, it swims into the unlighted region.

Many of the individuals react as soon as the anterior end reaches the lighted area, so that less than one-fourth of the body is in the light. This shows that light falling upon the anterior end alone is sufficient to cause the reaction.

A few specimens swim completely into the lighted area, then react

*Fig. 11.—Method of studying the manner in which Stentor reacts to light. The figure shows a circular area, illuminated from below, with the light cut off from the left side by a dark screen, the line *x-y* separating the light from the dark area. The Stentors collect in the dark area. The reaction of a specimen which comes to the line *x-y* is shown at 1, 2, 3, 4.

in the manner above described. In such cases the nature of the reaction is seen with especial clearness, the entire animal being projected on the screen and the differentiations of bodily structure (mouth, oral and aboral sides, etc.) being conspicuous. Specimens which swim completely into the lighted area are usually compelled to react two or more times before they escape from the lighted region.

When the light is cut off entirely the Stentors distribute themselves throughout the dish. If the light is now admitted from below, the unattached Stentors in the lighted area react by swimming backwards a certain distance, turning toward the right aboral side, then swimming forward again. This reaction is repeated frequently until after an interval the Stentors are carried by these movements outside the lighted area. They then cease to give the reaction. The reaction, under these conditions, is thus the same as that produced when Stentors or Paramecia are subjected to other adequate stimuli, as when they are placed in a chemical or dropped into very warm or very cold water. The result of the reaction is, in every case, to remove the organism from the sphere of action of the stimulus. When the stimulus is light this result is produced in exactly the same way as when the stimulus is heat or cold or a chemical.

The same results may be obtained by lighting the vessel containing the Stentors directly from above and shading one portion with a screen. The Stentors remain in the shaded region, responding by the motor reaction above described when they come to the lighted area. With a favorable culture the experiment succeeds even when the source of light is comparatively feeble, as when an ordinary incandescent electric light is used as the source of illumination.

The results so far show that a sudden increase in the intensity of illumination induces in Stentor a reaction which is of the same character as the reaction to other strong stimuli. Such a sudden increase may be due either to the passage of the Stentor from a dark to a light region, or to a sudden increase in the brightness of the light which falls upon the animal. The general effect of the reaction is to prevent the Stentor from entering a brightly illuminated area, or to remove it from such an area.

We may now arrange the conditions so that the light shall come from one side, while at the same time differences in illumination shall exist in different regions. This may be done by illuminating the vessel containing the Stentors from the side, then covering one portion of the vessel with a screen.

The organisms are placed before a lighted window, or an incandescent electric light, in a vessel with a plane front (Fig. 12). One-half of the vessel is then cut off from the light by a screen (s), the shadow of which passes across the middle of the vessel containing the Stentors. One side of the vessel is thus in the light, the other in the shadow, and these two regions are separated by a sharp line (Fig. 12, $x\cdot y$).

The Stentors are soon all collected in the shaded side of the vessel. Here they swim about freely in all directions, but do not cross the line into the lighted portion. Now, by focusing the Braus-Drüner on this line, the behavior of the individuals on reaching it may be observed.

It is well to examine the conditions in this case with care, as they present opportunities for a precise and crucial test of the theory that the reaction to light is due to a direct orientation through the falling of light on one side of the organism (phototaxis or phototropism in the strict sense, as defined by Holt & Lee). In the lighted portion of the vessel the rays of light come from a certain direction, as indicated by the large arrows (Fig. 12). In the shaded region there is not enough light to produce orientation, the animals swimming in every direction. On passing from the shaded region across the line $x\cdot y$ into the lighted region, the animal should (according to the tropism theory) become oriented. According to the theory of negative phototaxis by direct orientation due to differential action on

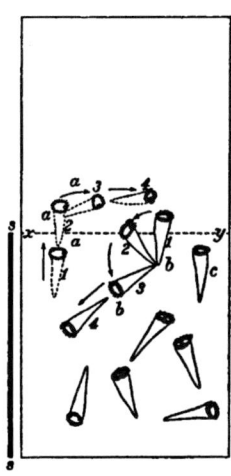

FIG. 12.*

* FIG. 12.—Method of testing the manner of reaction to light in Stentor. The large arrows show the direction from which the light rays come. A screen (s) cuts off the light from half the vessel, leaving a line ($x\cdot y$) separating a shaded part from a lighted part. The Stentors collect in the shaded part, here swimming about without orientation. At a (1, 2, 3, 4) we see a diagram of the reaction required by the tropism schema when the organism swims across the line $x\cdot y$, while at b (1, 2, 3, 4) we have a diagram of the reaction as actually given under these conditions.

the two sides, the animals on crossing the line should become oriented by turning *directly* away from the source of light, as shown in the diagram (Fig. 12) at *a*. The animal would then be expected to swim in the direction *x-y* as shown by the specimen *a*, 1, 2, 3, 4.

It cannot be held that the real source of light for the Stentors is that reflected from the bottom or sides of the dish in the lighted region, and hence coming on the whole from a direction perpendicular to the line *xy*, for *the behavior of the Stentors shows that this is not the case*. A Stentor in the shaded region, close to the line *x-y*, as at *c*, Fig. 12, receives whatever light there may be thus reflected exactly as it does after it has crossed the line, yet it shows no reaction and does not orient itself in any way. On the other hand, as soon as it crosses the line *x-y*, so as to receive the light coming from the window, it reacts strongly, as we shall see. It is thus clearly the light from the window, coming in the direction shown by the large arrows, that causes the reaction; hence the Stentor ought, according to the direct orientation theory, to orient itself in the line of these rays.

When a Stentor, swimming at random, reaches the line *x-y*, it reacts by stopping suddenly, then turning toward its aboral side, then swimming forward. It thus swims about until its anterior end is again within the shadow, where it continues to swim forward (Fig. 12, *b*, 1, 2, 3, 4). Often the first reaction is not sufficient to direct it into the shadow; in this case the reaction is repeated; one to three reactions almost invariably bring the Stentor back into the shadow. It has no particular orientation in the shadow, but swims in whatever direction it happens to be headed.

Very frequently the animals react when the anterior end alone has crossed the line, so that less than the anterior half of the body is lighted. In other cases the animal swims completely across the line, sometimes for a distance greater than its own length, into the light, before it reacts. In any case the reaction is that above described.

Does the Stentor, when it turns on entering the light, always turn away from the source of light, as the theory of direct orientation requires?

At the moment of crossing the line into the light the Stentor may occupy various positions. It will be well to note specifically the reaction in certain of these positions, as we obtain here the observations which furnish an exact and crucial test of the direct orientation theory.

1. The Stentor may reach the line with the aboral side directed toward the source of light (Fig 12, *b*). It therefore turns (as usual) toward its aboral side. It thus swings its anterior end *toward the source of light*, in the direction opposite that required by the direct

orientation theory. This observation was made repeatedly in a very large number of cases; not a single exception to it was observed. The swinging of the anterior end is continued past the point where the light falls directly upon it until the animal is directed again into the shadow, as illustrated in the diagram (Fig. 12, *b*, 1, 2, 3, 4).

2. The Stentor may reach the line with the aboral side directed away from the source of light. In this case it turns (as usual) toward the aboral side, thus swinging its anterior end *away* from the source of light.

3. The Stentor may reach the line with the aboral side directed upward or downward or in some intermediate position. In every case it turns toward the right aboral side, in whichever way this is directed.

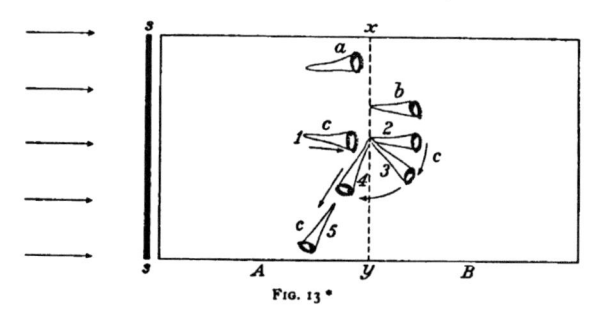

FIG. 13.*

The writer wishes it to be understood that the foregoing statements as to the direction in which the animal turns are presented, not merely as interpretations in accordance with a certain theory, but as direct, unequivocal observations, many times repeated. Thus, on passing from a darker to a lighter area, even when the light comes from one side, the Stentors react merely to the difference in illumination, without regard to the direction from which the light comes. The direction of turning is determined throughout by an internal factor, not by the side of the animal on which the light falls, nor by the direction of the rays of light. We have put the theory of orientation by direct differential

---

*FIG. 13.—Another method of testing the manner in which Stentor reacts to light. For a side view of this apparatus, see Fig. 14. Light comes from the left side, in the direction indicated by the arrows. A screen (*s*) is interposed between the source of light and the vessel containing the Stentors. This screen is of such a height (as illustrated in Fig. 14) that it cuts off the light from the half (*A*) of the vessel next to the window, leaving the other half (*B*) lighted. At *c* (1, 2, 3, 4, 5) is seen the reaction method of a specimen which swims across the line *x-y*, separating the shaded half *A* from the lighted half *B*.

action of the light on the two sides of the animal to a precise test, and found it to be incorrect.

The same result is brought out in perhaps a still more striking manner by the following method of experimentation: A vessel containing Stentors is placed on a dark background near a source of light (a window or an incandescent electric lamp). The light thus comes from one side and a little from above. An opaque screen is placed between the window and the vessel containing the Stentors, of such a size and in such a position that the top of the shadow of the screen falls across the middle of the vessel on the line x-y (Fig. 13; see also Fig. 14). Thus the half of the vessel next to the window (A) is darker than the farther half (B), and the Stentors collect in this shaded half. After some time scarcely a specimen is found in the lighted part of the vessel away from the window. The conditions in this case are illustrated in the side view (Fig. 14).

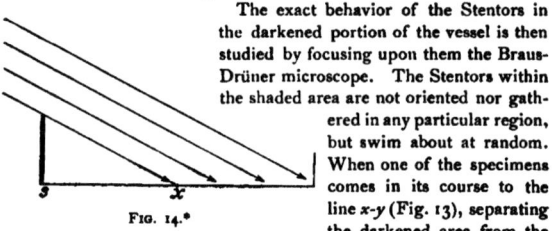

FIG. 14.*

The exact behavior of the Stentors in the darkened portion of the vessel is then studied by focusing upon them the Braus-Drüner microscope. The Stentors within the shaded area are not oriented nor gathered in any particular region, but swim about at random. When one of the specimens comes in its course to the line x-y (Fig. 13), separating the darkened area from the light, it responds to the sudden light which falls upon it from the window by giving the motor reaction, turning to the right aboral side and swimming back into the shaded region. Often the reaction occurs as soon as the anterior end of the Stentor has crossed the line x-y, so that the entire Stentor does not pass out into the lighted area. In other cases the specimen crosses the line x-y completely before the reaction occurs, so that the entire body is illuminated. It then reacts in the usual manner, turning toward the right aboral side, so that it is headed toward the shaded region; thus swimming back across the line (Fig. 13, c). After returning into the shaded region the animals swim about at random as before.

What is the reason for the return of the Stentor into the darkened area after it has crossed the line into the light region?

---

* FIG. 14.—Sectional view, from the side, of the conditions shown in Fig. 13. The arrows show the direction of the light rays. The region from s to n is shaded by the screen s.

By so doing it swims toward the window, thus in the direction from which the strongest light is coming. According to the theory of phototaxis as due to the direct action of the light on the motor organs of the animal, this movement is inexplicable. Thus, in the analysis of this theory given by Holt & Lee (1901), it is shown that in the case of a negative organism, such as Stentor, light of supraoptimal intensity, like that coming from the window, must be assumed to cause increased contraction of the cilia. After the organism has passed across the line *x-y*, or while it is passing across this line, it has the anterior end directed away from the source of light; according to the tropism theory this is a stable position and should not be changed. For, supposing the organism swerves a little toward either side, the cilia on that side will be more strongly affected by the light, so that the animal will at once be turned back into the position of equilibrium with anterior end directed away from the light.

Nevertheless, under these circumstances the organism *does* turn and swim back into the darkened area. An explanation for the apparent movement of a negative organism against the direction of the light rays is sometimes given in the following form: The light from the window is said to fall upon the side or end of the dish farthest from the window and is reflected back, so that the chief source of light for the Stentors is not the window, but the side of the dish opposite the window. The animal therefore becomes oriented with relation to this source of light and swims away from it.

Comparison of the movements of the Stentors in the darkened area *A* with those in the lighted area *B* shows that this explanation can not possibly be correct. Consider an individual at the point *b*, Fig. 13, which turns and swims toward the window into the dark region. It is affected by light from two sources, (1) from the window, (2) reflected from the side opposite the window. According to the above theory the turning is due to the fact that the light from the opposite side is of greater strength than that from the window (in itself a most improbable suggestion). Compare this Stentor *b* with an individual at *a*, in the darker region. This animal receives no direct rays from the window, yet does receive the reflected rays from the opposite side. If these reflected rays are sufficient to cause *b* to become oriented in spite of the opposing rays from the window, they must produce the same effect, *a fortiori*, on the individual *a*, since they are the only rays which reach it. Yet individuals in the position *a* do not become oriented at all. The individuals in the shaded portion of the vessel swim about in all directions, without relation to the direction of the light rays. It is only when they come to the line *x-y*, where they would pass into the region lighted directly from the window, that they react by turning toward the right aboral side and passing back into the shadow. It is thus clear that it is the light *coming from the window* to which they react, not the light reflected from the sides of the dish. We have here realized the condition concerning which there has been so much discussion, and which has been considered impossible and unrealizable by various authors—a negative organism reaching the darker region by swimming toward the source of strongest light.

This would of course be quite inexplicable on the tropism theory as set forth by Holt & Lee. What does it indicate as to the real nature of the reaction? To this inquiry there can be but one answer. The organism reacts on passing from a darker to a lighter area, without regard to the direction from which the light comes. It reacts to the increase in the amount of light falling upon it as compared with the condition an instant before it had passed into the lighted area. The reaction takes the usual form—a backing and turning toward the right aboral side, followed by a forward motion. The organism, therefore, is directed again toward the shaded area, which it enters.

In all our experiments thus far there have been marked differences in the illumination of different areas. Let us now arrange the conditions so that light comes from one side, and all parts of the vessel are equally illuminated. This may be done by placing the Stentors in a glass vessel with plane walls at one side of a source of light, such as a window or the bulb of an incandescent electric light. The Stentors, after a very short interval in which the reaction seems indefinite, swim away from the source of light, thus gathering at the side away from the window, where they move about in a disordered way. During the reaction the Stentors are *oriented*, with the longitudinal axis in the general direction of the light rays and with the anterior end away from the source of light.

Thus while it is true that the direction of the rays of light has little if any effect on the reaction when the animals are at the same time subjected to a sudden change from dark to light, it does determine the direction of movement when acting alone. In order to discover just how the reaction occurs it is necessary to observe the animals at the moment when they change from their former undirected swimming to the movement away from the source of light.

For determining this a large number of Stentors are placed in the dish next the window on a dark background. The light comes from one side and a little from above. The direct rays of the sun were not employed.

Above the glass vessel are focused the lenses of the Braus-Drüner stereoscopic binocular. This gives a magnification of 65 diameters, with a working distance of 3 cm., and permits exact observation of the movements of the individual Stentors. To one who has worked only with the monocular microscope, the use of the stereoscopic binocular in studying the movements of small organisms will be a revelation.

The vessel containing the Stentors is first covered with a dark screen and the Stentors are allowed to become equally distributed throughout the dish. The screen is then raised, allowing the light from the window to fall upon the Stentors. Those which are swimming in any other direction than away from the window now turn and in a short time are swimming toward the side of the dish away from the window.

With the Braus-Drüner the movements of individuals are observed at the moment of removing the screen. Some turn at once, while most continue for a few seconds in the direction in which they are swimming and then turn. *All turn in every case toward the right aboral side.* The turning is continued or repeated until the anterior end is directed

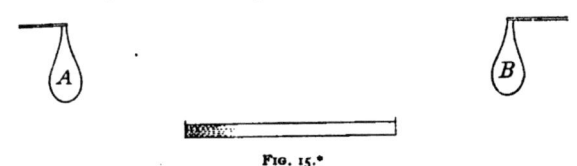

FIG. 15.*

away from the window; then the direct course is continued, carrying the Stentor to the side of the dish away from the window. *The direction of turning is thus determined by an internal factor*—the structure of the body.

The behavior of the Stentors may be controlled and studied more exactly by a different order of experimentation. The animals are placed in a shallow rectangular glass vessel on a dark background, in a room that is entirely dark save for two incandescent electric lights *A* and *B* (Fig. 15). These are clamped in position, one on each side of the dish containing the Stentors, and about eight inches from it. Both these lights can be turned on at once; both can be extinguished or one can be turned on while the other is turned off. When only one is turned on the direction of the light can be instantly reversed by simultaneously extinguishing this one and turning on the other.

With both lights extinguished the Stentors in the vessel are allowed to become equally distributed; then *B* is illuminated. In a short time

---

* FIG. 15.—Method of testing the reaction of Stentor to light. *A* and *B* are incandescent electric lights.

most of the individuals have gathered at the side next to A, as in Fig. 15. Then B is extinguished, while at the same time A is illuminated. The Stentors then turn and move toward B. They may be stopped at any point in their course and the direction of swimming reversed by simultaneously turning off one light and turning on the other. With a sensitive culture the phenomena take place with considerable precision, about four-fifths of the individuals responding quickly to every reversal of the direction from which the light comes.

Under these circumstances it is easy to observe the individuals at the moment of the reversal of the course. The observation already made is confirmed; the animals always turn at the moment of reversal toward the right aboral side. The reaction is thus of the same sort that occurs when there is a sudden increase in illumination. After the first reaction the anterior end is pointed in a new direction. If this new direction is away from the source of light the animal swims forward in the course so laid out. If, as is usually the case, the first reaction does not result in directing the anterior end away from the source of light, the reaction is repeated, and this may occur several times. Thus the anterior end becomes directed successively toward every quarter; as soon as it lies toward the side opposite the light the reaction ceases. The animal now swims straight ahead (that is, in a spiral with a straight axis) away from the source of light.

Thus while it is clear that light falling from one side produces a well-defined orientation, this orientation does not take place in such a way as to be in accordance with the tropism theory as set forth, for example, by Holt & Lee. It is not the direct action of the light on the motor organs of the side on which it impinges that determines the direction of turning, but the latter is due to an internal factor. This becomes still more evident when the conditions are so arranged that the direction of turning demanded by the internal factor is the opposite of that required by the tropism theory.

These conditions can be fulfilled in the following manner: The light to be turned on (Fig. 15) is so moved beforehand that its rays shall fall, not directly on the anterior end of the Stentor, but obliquely at an angle to the path they are following. The animals then react as before, by turning toward the right aboral side. It often happens that this involves first a direct turning toward the light, as illustrated in Fig. 16. In such a case the turning is continued or repeated until the anterior end is directed away from the source of light. We have seen the same result produced under similar conditions in the experiments illustrated in Fig. 13.

What is the real stimulus to the production of the motor reaction which results in orientation? The experiments directed precisely upon this point show that the stimulus producing the motor reaction is an increase in the intensity of light upon the sensitive anterior end. Now, in the reaction to a continuous light coming from one side, the conditions are present for exactly such changes in the intensity of light at the anterior end as would induce the observed reactions. In the spiral course the animal swerves successively in many directions. In certain directions the swerving subjects the anterior end to a more intense illumination. This change acts as a stimulus to produce the motor reaction, which carries the anterior end elsewhere. In other directions the swerving leads to a decrease in the intensity of light affecting the anterior end. In this case no reaction is produced, and the organism continues to swim in that general direction. The details of this method of reacting will be given in the account of the reactions of Euglena, where the matter was subjected to careful analytical experimentation. The evidence all indicates that the conditions in Stentor are exactly parallel to those in Euglena.

We may sum up our results on Stentor as follows: A change from dark to light, such as is caused by swimming from a shaded into an illuminated region, acts as a stimulus to produce a typical motor reaction; the Stentor backs and turns toward the right aboral side, so that it returns into the shaded region. A change in the illumination of the anterior end produces the same effect as a change in the illumination of the entire organism. The direction from which the light comes has no observable effect on this reaction. But when the illumination is uniform and the light comes from a definite direction, then light falling on the anterior end of the Stentor causes the reaction, while light falling upon the posterior end causes none. The result is that the animal turns (toward the right aboral side) until its anterior end is

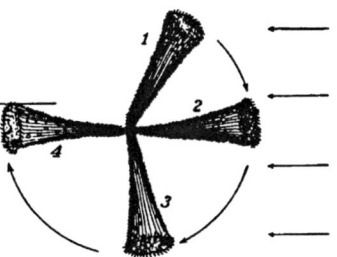

FIG. 16.*

*FIG. 16.—Method by which Stentor becomes oriented to light, when the light falls on the aboral side of the animal. Stentor turns, as shown by the arrows, at first toward the light, but the turning is repeated or continued until the anterior end is directed away from the light

directed away from the source of light, and swims in the direction so determined. The reaction to light is of essentially the same character as the reaction to other usual stimuli, and takes place by what we may call the method of trial and error. When the animal comes to the boundary of a lighted area, or when the anterior end is illuminated, this constitutes error; the animal tries some other direction, and repeats the trial till the condition constituting error disappears.

Are these results in agreement with all the observed facts? The only point on which perhaps question might arise is in regard to the production of a clearly marked orientation such as we find shown by Stentor when the light falls upon it from one side. In this case, as

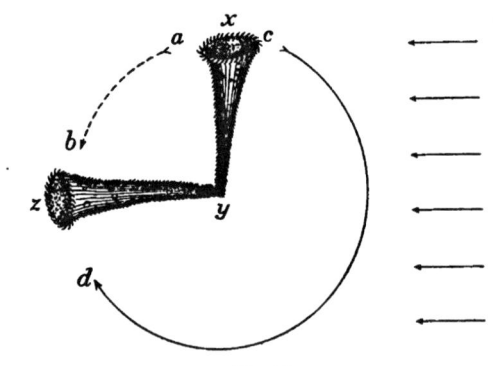

FIG. 17.*

we have seen, Stentor swims directly away from the source of light, and shows thus a typical orientation. As we have had the dictum that a motor reaction, such as I have described, "cannot account for an orientation" (Garrey, 1900, p. 313), it will be well to examine this matter a little farther. In a previous paper (Jennings, 1900, a) I have shown how orientation could be produced through a motor reaction; the case of Stentor exactly realizes the possibility there set forth. If

―――――――――
*FIG. 17.—Diagram to illustrate the difference between the method of orientation to light required by the tropism schema and that which actually takes place. To light coming from the direction shown by the straight arrows the tropism schema requires that an organism in the position x-y should attain the position y-z by turning in the direction indicated by the (broken) arrow a-b. The position is actually attained by turning in the direction indicated by the long arrow c-d.

the organism is at first not oriented to lines of influence coming from a certain direction, as in Fig. 17, *x-y*, and then becomes oriented, as at Fig. 17, *y-z*, there are clearly more ways than one by which the orientation can be produced. The essential question for deciding as to the nature of the reaction is not whether orientation occurs, but *how the orientation is brought about*. This consideration has been too often lost sight of in discussions of the behavior of the lower organisms.

According to the theory of tropisms, as defined by Verworn, Loeb, and Holt & Lee, the orientation should be brought about by the differential action of the external agent on the different sides of the organism; the organism should turn *directly* into the line of action of the external agent, and the direction of turning should be determined by an external factor, the direction of the infalling rays, or the side on which they strike the organism. Now this is a matter which can be settled by direct observation. Direct observation shows us in Stentor that orientation is not brought about in the manner demanded by the theory. The direction of turning is determined by internal factors. The reaction which produces orientation is identical with the typical reaction to a mechanical shock, to chemicals, to heat and cold. The difference between what is demanded by the theory of tropisms and what is actually observed may be made quickly evident to the eye by Fig. 17. According to the theory of tropisms the orientation of a negatively phototactic organism should take place by turning in the direction of the arrow *a-b*; in a Stentor in the position shown (*x-y*), orientation actually occurs by turning in the opposite direction, as shown by the arrow *c-d*.

The further question then arises as to why the organism remains oriented. All the facts point, in the case of Stentor, to the conclusion that the reaction to a constant light is due to the intense illumination on the sensitive anterior end. As soon, therefore, as the anterior end is turned away from the light, as is the case in the position *y-z*, Fig. 17, there is no further cause for reaction; the animal therefore remains with its anterior end directed away from the light; that is, it remains oriented. If, as a result of reaction to some other stimulus, or in any accidental manner, the animal comes into a position such that it is no longer oriented, the "motor reaction" is repeated until the animal comes again into the position of orientation in which it is no longer stimulated.

How does the method of reaction to light here described for Stentor agree with what we know of light reactions in other ciliates? As noted in the introductory paragraphs, comparatively little is known as to light reactions in this group of organisms. The observations of

Engelmann (1882, *a*) on the light reactions of certain green ciliates (*Paramecium bursaria*, *Stentor viridis*, etc.) were made before the typical motor reaction—the turning toward a certain structurally defined side—had been observed in any of the infusoria. Engelmann, therefore, paid no attention to this point. Yet there is much in his account of the reactions to light in these organisms to suggest that it takes place in a way similar to that which I have described above for *Stentor cæruleus*. Indeed, Engelmann's account, so far as it goes, fits precisely into the reaction method which I have described above. He found, as I have, that the organisms react either when only the anterior end is affected, or when the entire organism is flooded with light from beneath. The reaction consists in a sudden turn to one side, or a sudden start backward, just as in *Stentor cæruleus*. The only point which is lacking in Engelmann's account is the observation as to which side the organism turned; to this point he did not direct his attention.

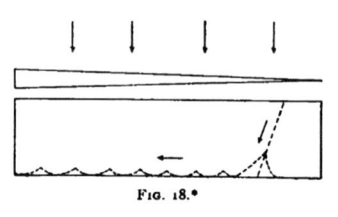

It is interesting to note that in the account given by Verworn (1889, Nachschrift) of the reaction to light in *Pleuronema chrysalis* there is nothing tending to support the theory of an orienting tropism. According to Verworn the reaction of Pleuronema to light is by a sudden leap (" Sprungbewegung "), which is repeated several times if the light continues. This sudden leap seems identical with the "motor reflex" which I have described as the typical reaction to stimuli in many ciliates, and which consists usually in a leap backward, followed by a turning toward a structurally defined side. It is in this manner, as we have seen, that *Stentor cæruleus* reacts to light and the reaction, as in Pleuronema, is often repeated many times.

FIG. 18.*

Thus the other carefully studied accounts of reaction to light in the Ciliata, while incomplete, agree so far as they go with that which I have given for Stentor, and contain nothing to suggest the idea of an orienting tropism dependent upon unequal stimulation of the motor organs on the opposite sides of the animal.

*FIG. 18.—Diagram of the reaction of Stentor to light, after Holt & Lee. Stentors are confined in a vessel behind a wedge-shaped prism containing a substance which partly cuts off the light, so that one end of the vessel is darker than the other. The usual course of a Stentor near the lighter end is shown by the broken line.

Davenport's reference (Davenport, 1897, p. 189) to the negative light reaction of Stentor makes no attempt to explain the mechanism of the reaction. Holt & Lee (1901) have given an account of some features of the light reaction of *Stentor cæruleus*. They did not attempt to determine directly the mechanism of the reactions, by observation of the exact movements of the organism. Specifically, they made no observations to determine whether Stentor becomes oriented by turning directly away from the source of light or only indirectly through a "motor reaction" such as I have described. They did attempt, however, to show that the gross phenomena observed might be interpreted in accordance with the prevailing theory of tropisms set forth on page 7 of the present volume. It will be well, therefore, to examine their observations in order to determine whether they contain anything inconsistent with the account set forth in the present paper.

Holt & Lee studied the behavior of Stentor in an elongated trough which was lighted from one side. The light passed through a prism which contained a translucent fluid (a weak solution of India ink), by means of which a portion of the light was cut out (Figs. 18 and 19).

At the thicker end of the prism more light was cut out, hence this end of the trough (Fig. 19, D) was darker than the opposite end (L). It was found that when Stentors were placed in the trough close behind the prism (at o, Fig. 19) they turned and swam away from the lighted side till the back of the trough was reached (a to d, Fig. 19). This is of course exactly what happens when no prism is interposed. Reaching the back of the trough the animals give the motor reaction (by backing, then turning toward the right aboral side), thus coming into either the position e or the position f (Fig. 19). They then swim forward again, strike the wall,

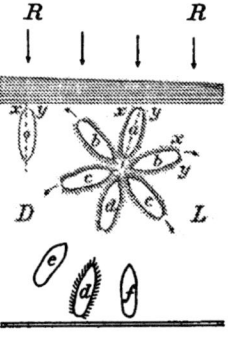

FIG. 19.*

* FIG. 19.—Reaction of such an infusorian as Stentor to light, under the conditions shown in Fig. 18. After Holt & Lee. The animal in the position x-y, close behind the prism, turns and swims to the position d, where it comes against the rear wall of the trough. It then turns either into the position e, toward the darker end D, or into the position f, toward the lighter end L. In the latter case it usually soon reacts again, and by repetition of the reaction it finally, as a rule, becomes directed toward D. Thus, finally, most of the Stentors collect in the dark end of the trough.

and repeat the reaction. This is repeated many times until the organisms are swimming either toward the end D or toward the end L. In course of time it is found that the preponderance of movement is toward the dark end D, so that the majority of the Stentors are gathered at D. Why this should be so is explained by Holt & Lee as follows:

> The reason why the Stentors went eventually in greater numbers toward D, and thus appeared oftener to choose e than f, is that such Stentors as went to e progressed farther toward D than those which went to f could progress toward L. These latter would soon strike the wall a second time, now pretty nearly at right angles, and during the recoil the light stimuli would favor a return to d. It appears then amply possible that the circumstance that the organism encounters the wall of the trough at an acute angle is sufficient to cause its farther progress to be, in the long run, toward D.

There is evidently nothing in this account which is inconsistent with the method of light reaction which I have described. On the contrary, the reason why the organisms finally swim toward the dark end and gather there becomes much more evident when the reaction method that I have described is taken into consideration. Let us suppose that the Stentors, after striking the back of the trough, turn in equal numbers toward D and toward L. In those swimming toward D the anterior end is directed away from the source of strongest light (due to reflection from the lighted end of the dish L), and the animals are passing into a region of less intense light. There is thus nothing to cause the "motor reaction," with its accompanying change in the direction of movement. In the Stentors swimming toward L, on the other hand, the strongest light falls on the anterior end, and the organisms are passing into a region of more intense light. Either of these factors taken separately may, as we have seen, cause the motor reaction (the turning toward the right aboral side), thus changing the direction in which the Stentors swim. The animals which start to swim toward L will therefore soon be turned, and only when the direction of movement is toward D will there be no cause for further change.

The observations of Holt & Lee are thus quite in harmony with the reaction method which I have described, and indeed receive illumination when this reaction method is taken into consideration.

In the "fourth case" discussed by Holt & Lee (*loc. cit.*, pp. 475–478), the two factors mentioned as determining the turning of the Stentors away from the end L would work in opposite directions; only experience can tell which would be more effective. As Holt & Lee do not state specially that they observed the reactions of Stentor under these conditions no comment is required. Experiments of this character will be further considered after we have described reactions to light in flagellates.

THE FLAGELLATA.

In the following pages we shall examine the method of reaction to light in the flagellates *Euglena viridis*, *Cryptomonas ovata*, and a species of Chlamydomonas.

EUGLENA VIRIDIS.

Euglena viridis swims in a spiral path, continually swerving toward that side which bears the larger " lip " and the eye, the so-called dorsal side (Fig. 20). Its motor reaction to most stimuli is by a sudden pronounced turning toward the dorsal side; that is, by swerving still farther toward the same side toward which it swerves in its normal swimming. Thus the direction of its path is changed (Jennings, 1900).

The general features of the reaction of Euglena to light have been well worked out by Englemann (1882, *a*) and Wager (1900). These authors show that Euglena collects in lighted regions. The organisms pass into a lighted area without reaction. But on coming to the outer boundary of such an area, where they would pass out into the dark, they react by turning round and passing back into the light. The collections of Euglenæ in lighted areas are thus brought about in much the same manner as the collections of Paramecia in regions containing a weak acid (Jennings, 1899). If diffuse light falls from one side on water containing Euglenæ, the organisms swim toward the source of light. But if strong sunlight falls upon them they swim away from the source of light.

Engelmann showed that the colorless anterior end is the part that is chiefly sensitive to variations of light. Often the organism in a lighted area, on reaching the edge, reacts by turning when only the colorless tip has passed into the darkness.

The precise method of reaction to light, the direction of turning in becoming oriented or in passing back into the lighted area, was not worked out by the authors named. To this point we shall direct our attention.

When a large number of Euglenæ are swimming toward the source of light, if the illumination is suddenly decreased in any way, they give

FIG. 20.[*]

---

[*] Fig. 20 shows the spiral path of Euglena in its ordinary swimming.

the typical motor reaction described in my previous paper as a response to other classes of stimuli (Jennings, 1900, p. 235). That is, they turn at once toward the dorsal side (that bearing the larger lip and the eye). This is very easily seen when the Euglenæ are mounted in the ordinary manner in a thin layer of water on a glass slide and observed with the microscope in the neighborhood of a window. If the hand is interposed between the slide and the window all the Euglenæ react in the way just described.

The reaction is a very sharp and striking one and produces a very peculiar impression. At first all the Euglenæ are swimming in parallel lines toward the window. As soon as the shadow of the hand falls on the slide the regularity is destroyed; every Euglena turns strongly and may seem to oscillate from side to side in the manner described later.

The turning is often preceded by a slight movement backward. This was not observed in the reactions to other stimuli (Jennings, 1900, p. 235), though it agrees with what we find in most other ciliates and flagellates. In Euglena the reaction to variations in the intensity of light seems more sharply defined than to most other stimuli. The fact that the turning is always toward the dorsal side is observable with the greatest ease. It is particularly evident when the organisms are confined to a thin layer of water, so that they cannot swerve up or down, but only to the right or left.

The reaction occurs whenever the light is suddenly decreased in any way. Certain different conditions under which it occurs deserve special mention. (1) As we have seen, the reaction occurs when a screen is brought between the organisms and the source of light toward which they are swimming. (2) It also occurs when the illumination is decreased by cutting off light from some other source than that toward which they are swimming. Thus the organisms on the stage of the microscope may be lighted from below, by the substage mirror, and at the same time may receive light from the window at one side of the preparation. They swim toward the window, since the light from that quarter is much stronger than that from below. If now the light from below is suddenly decreased by closing the iris diaphragm, the Euglenæ react as usual by turning strongly. This is notwithstanding the fact that the proportion of light coming from the window, to which they were oriented, is now greater than before, so that it might be supposed that they would remain more strongly oriented than ever. For the rest, the disturbed orientation is soon restored. (3) The reaction occurs when the decrease in illumination is due to the movements of the Euglenæ; that is, when the swimming organisms come to the edge of a lighted region where they would, if the course were continued, pass into the darkness. As a result of the reaction they return into the light.

The reaction occurs at a decrease in illumination not only when the organisms are oriented and swimming toward the source of light, but also when they are not oriented and are merely scattered in a weakly lighted area. Further, in cases where most of the Euglenæ are oriented and swimming toward a source of light, a number of specimens will always be found that are not oriented at all, or are swimming away from the source of light. Such individuals react to a sudden decrease in illumination in the same manner as do the specimens that are oriented with the anterior end toward the source of light. This result may be observed in a curious way as a consequence of the fact that it requires some time for the light to produce its orienting effect. Thus, if the Euglenæ are placed between a weak and a strong light they swim toward the strong light. If, now, the strong light is cut off, they react in the usual way and swim toward the weak light. Now the strong light may be restored ; the Euglenæ continue for a few seconds to swim toward the weak light, thus away from the strong light. If while they are swimming in this manner the strong light is cut off, the Euglenæ, swimming away from it, react in the usual manner, by turning strongly toward the dorsal side.

The usual reaction may be produced by a decrease in illumination that is not sufficient to cause a permanent change in orientation. Thus the Euglenæ on a slide or in a shallow dish may be lighted from a window at one side. By passing a small screen in front of the window at some distance from the preparation a portion of the light is cut off; the Euglenæ then respond in the usual way, by swerving toward the dorsal side. The movement thus becomes very irregular. Since the Euglenæ continue to revolve on their long axes the dorsal side may lie first to the (observer's) right, then to the left. The Euglenæ all seem, therefore, to vibrate from side to side. This is the "Erschütterung" or trembling described by Strasburger (1878) as occurring in swarm-spores when the illumination is changed; it will be understood better when we have considered more in detail the mechanism of the reactions. Meanwhile the screen retains its position, but still admits more light from the direction of the window than from any other direction. The reaction of the Euglenæ, therefore, soon ceases; their orientation is restored in the way to be described later, and they continue to swim toward the window.

This experiment is an important one. It shows that the typical reaction may be produced by a decrease in light that is not sufficient to permanently destroy the orientation. Thus it is clearly the decrease in illumination to which the organisms react; not to a change in the direction of the light rays. The experiment shows further that it is not the absolute amount of light that determines the reaction. Some time after the decrease in illumination takes place the organisms behave just as they did before, swimming in the same direction. Further, the illumination may be decreased very slowly to the same extent without causing a reaction. If the screen is at first far away from the preparation and is then slowly moved to the position it occupied in the experiment just described no reaction is produced. It is only the sudden change that has caused the reaction. The change, however, need not be a very marked one in order to be effective.

Our experiments thus far have shown that in a moderate light Euglena reacts to a decrease in illumination. But the absolute amount of light present has an effect on the reaction. If the light is very strongly increased the same reaction is produced as when the light is decreased. If while the organisms are swimming toward a moderately lighted window direct sunlight is allowed to fall upon them, they respond in the same way as to a sudden decrease in illumination; that is, they turn strongly toward the dorsal side, continuing or repeating the reaction till the anterior end is directed away from the source of light. They now continue to swim in that direction, the positive reaction having been transformed into a negative one. Thus under intense light the conditions of stimulation are the opposite of those under moderate light. This is paralleled in the reactions of the infusoria to chemicals; often a strong solution of a certain chemical produces a reaction under opposite conditions from those in which a weak solution of the same chemical is effective.

Let us now proceed to a more careful study of the reaction itself. The reaction which occurs when the illumination is changed is really an accentuation of a certain feature of the usual movements. Euglena, as we know, revolves on its long axis as it swims forward, and at the same time it swerves toward the dorsal side. The resulting path is therefore a spiral one (Fig. 20). The usual reaction to a stimulus is an accentuation of this normal swerving toward the dorsal side, as compared with the other factors in the swimming; the organism suddenly swerves so much farther than usual in this direction that the path may be completely changed. If the reaction is a very decided one the revolution on the long axis and the movement forward may cease during the swerving toward the dorsal side; the anterior end then describes the arc of a circle about the posterior end as a center. In a less pronounced reaction the revolution on the long axis continues. The circle described by the anterior end is then less and the whole body describes the surface of a cone, or a frustum of a cone, as illustrated in Fig. 21. Every gradation exists between the normal spiral course and the strong reaction in which the anterior end swings in a circle about the posterior end as a center.

When oriented and swimming toward the source of light the swerving toward the dorsal side is comparatively slight. As seen from above, the organisms seem merely to oscillate a very little from side to side as they revolve on the long axis. Careful examination shows that the swerving is always toward the dorsal side, as in Fig. 20, the alternations in direction being due to the alternations of position of the dorsal side. Now, when the illumination is suddenly decreased, the Euglenæ at once swing much farther than usual toward the side to

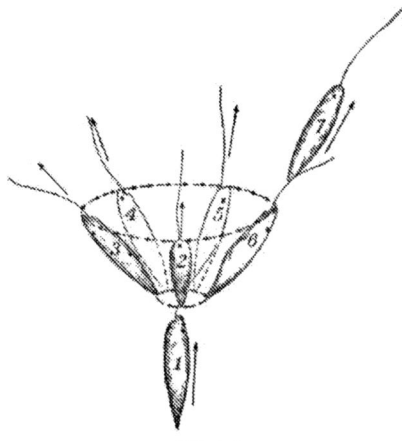

FIG. 21.*

which they are already swerving, that is, toward the dorsal side. If the decrease in illumination is not very great, so that the stimulus is not a strong one, the swerving is not very great, and the organism at the same time continues to revolve on its long axis; thus the anterior end describes a circle and the whole body describes the surface of a

* FIG. 21.—Diagram to illustrate reaction of Euglena when the illumination is decreased. The Euglena is swimming forward at 1; when it reaches the position 2 the illumination is decreased. Thereupon the organism swerves strongly toward the dorsal side. This swerving, combined with the revolution on the long axis, causes the anterior end to swing about a circle, so that the Euglena occupies successively the positions 2, 3, 4, 5, 6, etc. From any of these positions it may start forward, as indicated by the arrows, if the condition causing the reaction ceases to act. In the figure the Euglena is represented as swimming forward from the position 6.

cone, or the frustum of a cone, as indicated in Fig. 21. The result, as seen from above, is that all the specimens seem to vibrate from side to side; in other words, they are taken with a sudden oscillation or trembling. This oscillation when the intensity of the light is suddenly changed was observed by Strasburger (1878, pp. 25 and 50) in flagellate swarm-spores; he speaks of it as "Erschütterung" or "Zittern." During this oscillation the anterior end becomes pointed successively in many different directions, as Fig. 21 shows. When, now, the usual forward course is resumed (with only the usual amount of swerving toward the dorsal side), the animal follows one of these directions. Thus its path is changed (Fig. 22). Strasburger (1878, p. 25) noticed that the path followed after the oscillation was oblique to the former path. As a study of Figs. 21 and 22 will show, this is a necessary consequence of the increased swerving toward the dorsal side, to which the oscillation itself is due. All these relations become much clearer if a model of an actual spiral is studied; it is difficult to represent them upon a plane surface.

If the stimulus is stronger, as when there is a greater decrease in illumination, the swerving toward the dorsal side is much greater; the organism wheels far to that side, so that the spiral course seems entirely interrupted. But there is really nothing in this reaction differing in principle from what is happening in the normal forward swimming. If the swerving toward the dorsal side is long continued the specimen may be seen to swing first far to the (observer's) right, then, after it has revolved on the long axis, far to the (observer's) left; in reality it swings an equal amount upward and downward and in intermediate directions. It may, however, swing at once so far to the dorsal side that the new

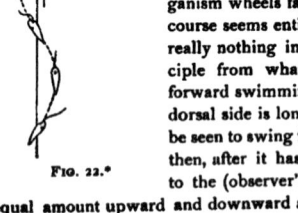

* FIG. 22.—Shows the spiral path of Euglena, illustrating the effect of a slightly marked reaction. At *a* the illumination is decreased; the organism therefore swerves toward the dorsal side, causing the spiral to become wider. At *b* the ordinary method of swimming is resumed; since at this point the organism was more inclined to the axis of the spiral than before the reaction, the new course lies at an angle with the previous one. Compare with Fig. 21.

course forms a right angle, or a still greater angle, with the original course; if the turning is through 180°, the course will be squarely reversed. Indeed, sometimes the organism swings around an entire circle or more. When the usual method of swimming is resumed after such reactions as those just described, the course has been completely changed.

Strasburger (1878, p. 25) noticed that after a decrease in illumination flagellate swarm-spores often turn strongly to one side or even describe circles. But he did not notice that the turning was always toward the same side of the organism,* and did not perceive the relation between this effect and the remainder of the reaction.

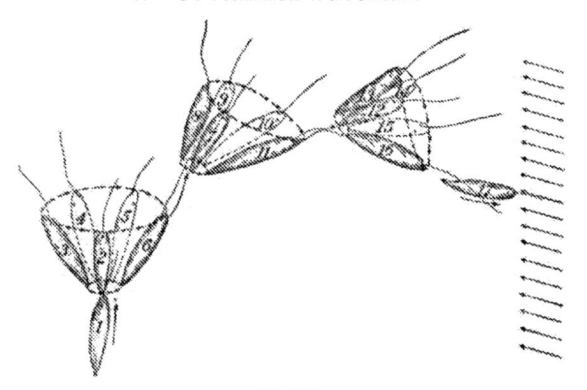

FIG. 23.†

This method of reaction is particularly striking when the Euglenæ are confined to a very thin layer of water between the slide and the cover glass, so that they cannot swerve up or down. When the light is decreased, we will suppose that the dorsal side is to the (observer's) left. The Euglena then swings far to the left. At the same time it

---

*Naegeli (1860, p. 96) had, however, before Strasburger, observed that in such swarm-spores the same side always faces the outside of the spiral path. This observation, which really contained the germ of a correct understanding of the reactions to stimuli, seems hardly to have been noticed by later writers.

† FIG. 23.—Diagram of the method by which Euglena becomes oriented with anterior end toward the source of light. At 1 the Euglena is swimming toward the source of light. When it reaches the position 2 the light is changed so as to come in the direction indicated by the arrows at the right. As a consequence of the decrease in illumination of the anterior end thus caused, the organism

revolves on its long axis, bringing the dorsal side down. Since it can not swing downward, owing to the narrow space, this has little effect on the reaction, save to stop the movement to the left. Now, by continued rotation the dorsal side has come to lie to the (observer's) right; the Euglena may then be seen to swing far to the right. In each case under these conditions it is at once evident by observing the larger lip at the anterior end that the organism is swinging toward the dorsal side.

This method of reaction is very effective in preventing Euglena from passing from an illuminated region to a shaded one. As soon as the anterior end enters the shadow, the animal swings far toward the dorsal side till the anterior end is brought again into the light, repeating the reaction if necessary. There is then no further cause for reaction. The reaction to a very strong increase of illumination is, as we have seen, identical with that to a decrease in illumination.

In our experiments thus far we have directed attention primarily to the effects of changes in the intensity of illumination, and have found that such changes produce a motor reaction independently of the direction of the light rays. But it is of course well known that Euglena does react with reference to the direction of the light rays. Euglenæ swim toward the source of light when weakly illuminated, away from the source of light when strongly illuminated. If Euglenæ are swimming at random in a diffuse light they soon become oriented when the light is allowed to act on them from one side, even if the intensity of illumination remains the same. Or, if Euglenæ are swimming toward a source of very weak light and a stronger light is allowed to act upon them from the opposite side, they become oriented, in time, with anterior ends toward the stronger light. In examining this dependence of the direction of swimming on the direction of the rays of light, we

---

swerves strongly toward the dorsal side, at the same time continuing to revolve on the long axis. It thus occupies successively the positions 2, 3, 4, 5, 6. In passing from 3 to 6 the illumination of the anterior end is increased; hence the reaction nearly or quite ceases. In the next phase of the spiral, therefore, the organism swerves but a little toward the dorsal side—from 7 to 8. But this movement causes a decrease in the illumination of the anterior end, and this change induces again the strong swerving toward the dorsal side. Hence in the next phase of the spiral the organism swings through 9 and 10 to 11. In this movement again the illumination of the anterior end is increased; hence the reaction ceases, so that from 12 the organism swerves only as far as 13. Then owing to the decrease in illumination caused by this movement, the swerving increases, so that the Euglena swings from 13 through 14 and 15 to 16. Now it is directed toward the source of light, and such swerving as takes place in the spiral course neither increases nor decreases the illumination of the anterior end. Hence there is no further reaction; the Euglena continues to swim forward in the direction 16–17.

shall have to keep in mind two questions: First, how is the position of orientation brought about? Second, what is the real stimulus in producing orientation?

To answer the first question we must observe the movements of the

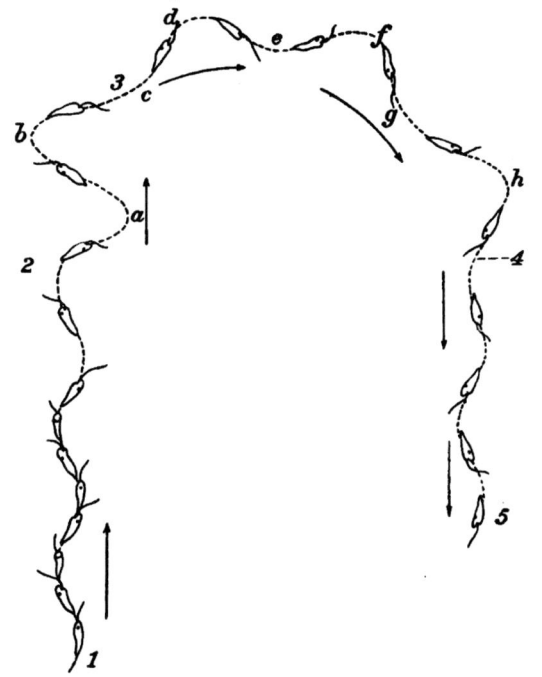

FIG. 24.*

organism at the time orientation occurs. Observation of the individuals as they are becoming oriented shows that orientation is brought about through the same motor reaction that we have already described;

---

* FIG 24.—Path followed by Euglena when the direction of the light is changed. From 1 to 2 the organism swims forward in the usual spiral path. At 2 the position of the source of light is changed, so that it now comes from behind. The organism then begins to swerve farther than usual toward the

that is, by a turning toward the dorsal side. The simplest case is perhaps that of the reversal of orientation, produced when strong sunlight is allowed to fall from in front upon specimens that are swimming toward a diffusely lighted window. Under these circumstances, as we have seen, the Euglenæ turn toward the dorsal side, changing their course. They may turn directly through 180°, in which case they are at once oriented with anterior ends away from the light; but usually the orientation is less direct than this. The reaction is generally repeated several times. Through its continued swerving toward the dorsal side, combined with the revolution on the long axis, the organism directs its anterior end successively in every direction. When the anterior end has finally come into a position where it points away from the strong light the reaction ceases, and the organism swims forward in the usual way. The details of the orienting reaction will be brought out more fully in the following account of the way in which the anterior end becomes directed toward a source of light of moderate intensity.

Let us now take a case in which the change in the direction of the rays of light is not accompanied by a change in the intensity of illumination. Euglenæ are swimming about at random in a diffuse light when all the light is allowed to fall upon them from one side. They then become oriented, with anterior ends directed toward the source of light. Or, the organisms are swimming toward a source of light when the direction of the light rays is changed or reversed by quickly moving the source from which the light comes. The Euglenæ then after a time become reoriented. Under such circumstances there is no sudden, decided reaction, such as occurs when the illumination is suddenly decreased. The organism merely begins to swerve farther toward the dorsal side than usual. Thus the spiral has become wider, and the anterior end comes to be pointed successively in many different directions, as illustrated at 1–6 in Fig. 23. In some of these positions the anterior end is directed farther away from the source of light, as at 3; in other positions more nearly toward the source of light, as at 6. In the latter case the swinging toward the dorsal side becomes less marked; hence the succeeding phase of the swing, which carries the anterior end away from the light, is less pronounced;

---

dorsal side, owing to the decrease in the illumination of the anterior end. Thus the spiral becomes wider, *a* and *b* showing the limits of the swerving. At 3 the normal amount of swerving is restored, so that the new path is at an angle with the old one. Now the organism swerves at each turn of the spiral a short distance away from the source of light, as at *c, e, g,* and a longer distance toward the source of light, as at *d, f, h,* for the reasons shown in Fig. 23. At *h* it has in this manner become directed toward the source of light, and there is no further cause for swerving more to one side than to the other; it therefore swims in a spiral with a straight axis toward the source of light.

the anterior end therefore does not swing so far in the direction away from the light as in the preceding phase it swung toward the light. This is illustrated at 7–8 in Fig. 23. But as a result of such swerving as does occur the anterior end is now (at 8) directed more away from the source of light than before. There then follows a new reaction, with increased swerving toward the dorsal side in the next phase of the spiral (8–11, Fig. 23), which carries the dorsal side toward the source of light. Hence the anterior end swings still further toward the position where the light shines directly upon it. This continues. As a result of this repeated swinging of the dorsal side slightly away from the source of light and strongly toward the source of light the organism gradually changes its course, continuing to swim in a spiral and to swerve toward the dorsal side, until the axis of the spiral is in line with the light rays and the anterior end is toward the source of light. This method of reaction will best be understood by a study of Figs. 23 and 24 and their explanation.

Thus the orientation is gradual and for a certain stretch after the light has begun to act the organism is not completely oriented. With a fairly strong light, however, the period of time required for complete orientation is very slight. Strasburger (1878, p. 24) noticed that when Hæmatococcus is swimming toward a source of weak light and the light is suddenly increased so as to reverse the orientation, there is a period of " verschiedenen Schwankungen " before the reverse orientation is attained. He paid little attention to the behavior of the organisms during this period, however.

Our account has been thus far purely descriptive; we have attempted to set forth the events as they may be observed, without trying to indicate the causes at work. We must now inquire as to what is the real stimulus and its method of action in producing orientation.

First, we note that in becoming oriented Euglena does not turn directly toward the source of light. As in the reaction to other stimuli, the turning is throughout toward a structurally defined side. This shows that the orientation of Euglena, like that of Stentor, cannot be accounted for on the orthodox tropism theory. In other words, the orientation is not due to the direct effect of the light on the motor organs of the side on which it falls. As in Stentor, orientation may be reached by turning either toward or away from the source of light, or in any intermediate direction. The response is a " motor reaction " of a definite type.

Just what is the stimulus which produces this motor reaction? All our experiments up to this point have shown clearly that this reaction is produced by changes in the intensity of illumination, and that a change in the illumination of the anterior end produces the reaction as well as does a change in the illumination of the entire body. Indeed, Engelmann (1882, a) showed that a change in illumination over the remainder of the body is ineffective in producing the reaction, so that in every case the reaction is due to the change in illumination at the anterior end. Now, in the orientation reaction the conditions are present for producing changes of illumination at the anterior end of precisely the character which would, in view of our other experimental results, bring about the reactions observed. This will best be shown by again examining in detail from this point of view a concrete case.

In Fig. 23 we will suppose that the Euglena at 1 is at first swimming toward the source of light. When it reaches the position 2 the light is changed, so that it now comes from the direction indicated by the arrows at the right. By this change the intensity of illumination at the anterior end is decreased, since before the light came from directly in front and affected the entire end, while now it falls upon but one side. We know from other experiments that as a result of such a change the organism reacts by swerving more toward the dorsal side, at the same time continuing to revolve on the long axis. This is exactly what happens now; by the increased swerving the organism is carried from position 2 to position 3. In this change the anterior end, swinging still farther away from the source of light, is still less illuminated than before. As a result of this farther decrease in illumination the reaction is continued or increased; combined with the revolution on the long axis it carries the organism successively to positions 4, 5 and 6. In this part of the movement the anterior end becomes pointed more directly toward the source of light, and is hence more strongly illuminated; there is therefore nothing in this movement to cause a reaction. The strong swerving toward the dorsal side then ceases or becomes less. But in the next phase of the spiral course (from 7 to 8), there is necessarily at least the normal amount of swerving toward the dorsal side, and this carries the organism to a position (8), where the intensity of the light acting on the anterior end is decreased. As a result of this decrease we know that the "motor reaction" must again be induced; the organism swings then farther toward the dorsal side· This movement, combined with the revolution on the long axis, carries the Euglena through 9 and 10 to 11. Here again the swerving decreases, because the change was from a less illuminated to a more illuminated region. Hence after reaching 12 the Euglena swerves only a little away from the light, to 13; then, as a result of the decrease in illumination at the anterior end caused by this movement, it swerves far toward the light, through 14 and 15 to 16. This movement causing greater illumination, the reaction ceases. The light is now shining full on the anterior end. The organism therefore swims forward in the usual spiral course, in all phases of which the illumination of the anterior end is equal. If the light came from the rear of Euglena 1 instead of from the direction indicated by the arrows, the reaction above described would be continued in the same way until the direction of swimming was completely reversed.

Thus the orientation of Euglena in a continuous light is due to the production of the "motor reaction," with its turning toward the dorsal side, whenever there is a decrease in illumination at the anterior end.

There is no other explanation of the orientation, so far as I am able to see, that is in agreement with all the facts. At first one is tempted merely to say that the subjection of the anterior end to shadow produces the motor reaction, and that this is continued until the anterior end is no longer shaded. This statement is correct if by "subjection to shadow" we mean an active process, involving a change from a more illuminated condition. But if we mean that darkness as a continuous, static condition is the cause of the reaction, then consideration shows that this will not account for all the facts. It leaves out of account the capability of the organism to become acclimatized to certain degrees of light and shade, and certain of the experimental results are crucial against it. Thus, suppose the Euglenæ are swimming toward a source of weak light, and a stronger light is then allowed to act upon them from another direction. The anterior end continues to receive the same amount of light as before (since the weak light still persists), yet the organism reacts as usual, becoming oriented toward the stronger light. The motor reaction by which the orientation is brought about cannot therefore be due to darkness or shade (considered statically) at the anterior end. On the other hand, the case just mentioned is easily understood on applying the explanation given above.

Again, it might be held that the reaction is due in some way to the *relative* amount of illumination at the two ends. It might be maintained, for example, that when the posterior end is more illuminated than the anterior, this difference acts as a stimulus to cause the "motor reaction." There is, of course, no independent evidence in favor of this view, and the experimental results prove it to be incorrect. We have shown that the reaction is produced (1) when both ends are equally stimulated, as when the light comes directly from one side; (2) when neither end receives light, as when the light is cut off completely. Further, it might be held that the reaction is produced when the anterior end is not more intensely illuminated than the posterior end. It is, of course, a little difficult to conceive how so indefinite a condition could act as a stimulus to a definite motor reaction, but in any case the experiments show that this is not the real cause of the "motor reaction." Thus certain of the experiments show that the "motor reaction" is produced even when the light is reduced by the same amount at both ends, so that the anterior end is still more strongly lighted than the posterior. This case is realized in the experiment in which a small screen is interposed between the Euglenæ and the window toward which they are swimming. The light is thus somewhat decreased, but is still sufficient to cause orientation. The anterior end is thus still lighted more than the posterior, yet the organisms respond with the "motor reaction" at the moment the light is decreased. The same thing is shown still more decidedly in the experiment described on page 50, in which the "motor reaction" is produced when the light is cut off from some other source than that toward which the organisms are swimming. In this case the proportion of light shining on the anterior end is greater after the change in illumination than before, yet the "motor reaction" is produced at the moment the change takes place.

The explanation we have given is, therefore, the only one that is in agreement with all the facts, and it accounts for every detail of the reactions to light. The cause of all the phenomena of light reaction in Euglena is the fact that a sudden change in light intensity on the anterior end induces a typical "motor reaction." It is noticeable that the reaction is throughout due to a dynamic factor, to some change in the relation of the organism to the light, a change due either to an active alteration of the environment, or to a movement of the organism. To static conditions, if not too intense, the organism may soon become acclimatized, so that no farther reaction is caused. The absolute intensity of the light affects the reaction only in so far as it determines whether it shall be an increase or a decrease in intensity that causes the "motor reaction."

To sum up, the reaction of Euglena, from beginning to end, is explained by the fact that a sudden change in illumination, even though slight, causes a definite motor reaction, the essential feature of which is an increased swerving toward the dorsal side. Orientation is brought about by the increased swerving in the next phase of the spiral course when the illumination of the anterior end is diminished, and by the decreased swerving in the next phase of the spiral when the illumination of the anterior end is increased. In general terms we can say that the reaction of Euglena to light is by the method of trial and error. The organism tries turning in many directions; when the turning is such as to produce a decrease in the illumination of the anterior end it "tries" other directions; when it is such as to produce increased illumination of the anterior end, or when no change in illumination results, the reaction ceases and the organism continues to swim forward in that position. The result of this method of reaction is necessarily orientation with the anterior end toward the source of light.

CRYPTOMONAS AND CHLAMYDOMONAS.

Cryptomonas ovata is one of the organisms studied by Strasburger (1878), under the name *Chilomonas ovata*, in his classical paper on reactions to light in flagellates and swarm-spores.

The specimens studied by the present author were mostly of the "young" form, having pointed, curved, posterior ends. One side is strongly convex, while the other is less curved, or is even concave near the posterior end. It is thus very easy to distinguish the two sides of the organism and to observe their relation to the movements.

Cryptomonas ovata swims in a rather wide spiral, with the more convex side toward the outer surface of the spiral. In other words, the organism swerves continually toward the more convex side. The response to usual stimuli is a strong turn toward this convex surface; this is easily seen when the organism comes in contact with an obstacle.

The Cryptomonads swim toward or away from the source of light under the same conditions as Euglena, and gather in lighted areas in the same manner as does the organism last named. They react to a sudden decrease in the intensity of illumination by turning toward the more convex side. If the decrease in intensity is marked, the organism turns suddenly for a long distance, 90° or more, so that the course is completely changed. If the stimulus is less the turning toward the more convex side is not so rapid, and since the revolution on the long axis is continued the body of the organism describes the surface of a wide cone or frustum of a cone. When a large number of specimens react in this way at the same time a peculiar shaking or trembling appearance is produced; this is evidently what Strasburger (1878) called " Erschütterung " or " Zittern." As a consequence of the wide swerving, when the normal method of swimming is resumed the course lies in a new direction.

In all these respects Cryptomonas exactly resembles Euglena. Further, the organism becomes oriented to light in precisely the same manner as is described above for Euglena. In fact, if we substitute " more convex side " for " dorsal side " in the account of Euglena, it will fit almost throughout the reactions of Cryptomonas. It is therefore unnecessary to describe the phenomena in Cryptomonas in detail.

A study was made also of the reactions of a species of Chlamydomonas. The movements of Chlamydomonas and its reactions to light resemble those of Euglena and Cryptomonas. But the organism is so small and the differentiations of the bodily structure are so slight that I was unable to determine the relation of its structure to the spiral path and to the direction of turning in the reaction. The oriented organism reacts to a decrease in illumination by a sudden turn to one side, by an increase in the width of the spiral, and by a change in the course, just as happens in Euglena and Cryptomonas. The unoriented organism becomes oriented in a manner which is similar to that described above for the two organisms just named. Since, however, I am unable to give the precise relations of these movements to structural differentiations of the body, a further account of details would not be of interest.

GENERAL RESULTS.

In summing up our results on reactions to light in the organisms studied, there are two points of especial interest which should be considered separately. The first relates to the nature of the reaction produced, the second to the nature of the agent causing the reaction.

NATURE OF REACTION PRODUCED BY LIGHT.

As to the nature of the reaction produced by light there has been much discussion. The orthodox tropism theory is perhaps that which has the greatest number of adherents. It is set forth in detail in the paper of Holt & Lee (1901). According to this theory the light acts directly on the motor organs of the side on which it impinges; supraoptimal light causes increase of the backward stroke (in the case of cilia or other swimming organs); suboptimal light causes a decrease in the backward stroke. The result is that the organism is turned directly toward or from the more intensely lighted side, and hence toward or from the source of light. The diagrams given in the preceding paper (Figs. 1 and 2) can be applied directly to the elucidation of this theory.

In the experiments on the ciliates and flagellates set forth in the present paper the precise method of reaction was determined by observation. It is not in accordance with the tropism theory above set forth. This has been emphasized in detail in the account of the reactions of Stentor, so that it need not be reiterated here. The reaction to light is of the same character as that to other stimuli, and takes the form of a motor reaction in which the organism performs a definite set of actions. It first usually stops or swims backward, then turns toward a structurally defined side, then continues forward. The result is to change the course of the organism. As a result of the continual rotation on the long axis, together with the swerving toward a certain side, the organism comes to be pointed successively in every direction. In continues to swim forward in that direction which does not induce a stimulus to further swerving. The whole reaction is a strongly marked example of the type of behavior which may be called the " method of trial and error."

NATURE OF AGENT CAUSING THE REACTION.

(1) The primary and essential cause of the reaction is a change of illumination. The change of illumination must take place with some suddenness, but need not be very great in amount. The *change* in illumination acts as an effective stimulus even though the degree of illumination preceding the change and that following it would, when acting continuously, produce no such result. This is shown by the experiments on Euglena, in which the light coming from one side was decreased a certain amount. The orientation of the organisms and their direction of movement was the same before and after the change, but at the moment the change occurred there was a marked reaction. Other experiments detailed above demonstrate the same thing. Further, the change in illumination acts independently of the direction of the rays of light. This is shown by the experiment just cited, in which the effective direction of the rays of light was the same before and after the reaction; it is also shown in the reaction caused when the light is decreased from below, in the case of Euglenæ swimming toward a window (p. 50), and in the reaction of Stentor on passing from a shadow to a lighted region even when the animal is oriented with anterior end away from the light (p. 39). The change in illumination acts equally whether it affects the entire organism or only the anterior end. The evidence indicates that in all cases it is really the change at the anterior end which induces the reaction.

(2) The absolute intensity of the light affects the reaction by determining in a given case whether a reaction shall be caused by an increase or a decrease in illumination. Through this action it also determines, in the way to be mentioned in the next paragraph, whether in a continuous light the sensitive anterior end shall be directed toward or away from the source of light; that is, whether the response shall be "positive" or "negative."

(3) Indirectly, and through the factor set forth in paragraph (1), the direction from which the light comes is a determining factor in the reactions. Through the spiral course in which the organisms swim such conditions are furnished that in a field continuously lighted from one side the sensitive anterior end of the unoriented organism is subjected to repeated changes in the intensity of illumination. As a result, organisms which respond by the motor reaction to an increase in illumination at the anterior end must become oriented with anterior end directed away from the light; organisms which react to a decrease in illumination must become oriented with anterior end directed toward the light. (Details in the account of Euglena, pp. 60, 61, and Figs. 23, 24.)

The results of this method of reacting may be stated correctly, though not completely, as follows: In a negative organism light falling upon the sensitive anterior end causes a reaction by which the anterior end is pointed in many different directions; the reaction ceases as soon as a direction is reached in which the anterior end is pointed away from the light. In a positive organism the shading of the sensitive anterior end produces the reaction by which the anterior end is pointed in many different directions; the reaction ceases as soon as the anterior end is no longer shaded.* The reaction is thus by the method of trial and error; when stimulated the organism tries many different positions, till one is found in which there is no further stimulation.

Consideration will show, I think, that the factors producing reaction to light in these lowest organisms are essentially the same as in higher ones, if man may be taken as a type of the latter. The factors are, as we have seen, variations in intensity of illumination, and, indirectly, the direction from which the light comes. It is possible that in man the latter factor works more directly than in the infusoria; leaving this question out of consideration, the two factors are present in both cases. Consider a human being who reacts to light as a purely physical agent, not with regard to the associations which it brings up. In a dark space a gleam of light is pleasant and induces movement toward it. There is then a positive reaction with orientation, but the orientation is not due to the difference in intensity of light on different parts of the body, nor to its direct effect on the motor organs. The orientation is such as to keep the light shining on the more sensitive part of the body, the eyes. An excessively powerful light is unpleasant and induces a negative reaction just as happens in Euglena; the orientation is then such as to keep the more sensitive part of the body, the eyes, away from the light. Further, man is sensitive to a sudden change in illumination. A strong light bursting from the darkness, or sudden darkness in the midst of bright light, induces a marked motor reaction, and less striking differences may produce a response. Both in man and in Euglena the reaction likewise depends upon color; but with this phase of the matter we are not at present concerned.

When the factors above set forth are taken into consideration certain peculiar experimental results that have given rise to much discussion become clearly intelligible. I refer particularly to the experiments in which the direction of the light and the decrease in intensity of illumination do not show the usual relations. Under ordinary conditions movement away from a source of light is movement into a region of less

---

* This statement is incomplete in that it does not bring out the fact that it is a *change* from light to shade or *vice versa* that induces the reaction; if this be understood, the statement is correct.

intensity; movement toward the light into a region of greater intensity. In the well-known experiments of Strasburger (1878) and others, this condition is modified by passing the light through a wedge-shaped prism filled with a solution that cuts out part of the light.

When a drop of water or a culture dish is placed beneath such a prism, and the latter is so situated that its surface is perpendicular to the light rays, the intensity of the illumination is greatest behind the thin edge of the prism, and thence decreases gradually toward the opposite end, while the rays of light all come directly from above. Under these conditions Strasburger (1878, p. 36) found that the positive swarm-spores remained equally distributed throughout the drop, not collecting at the lighter end. Now, the only difference between this experiment and the one illustrated in Fig. 11 of the present paper is that in Strasburger's experiment the decrease in illumination is very gradual. We have seen above (p. 52) that a very gradual change in illumination produces no reaction. Hence the organisms may wander from one side of the drop to the other without reaction, the difference in illumination at two successive instants never rising to the necessary threshold of stimulation. If the relation of stimulus to reaction follows Weber's law, the result is just what we should expect, provided the change in illumination is sufficiently gradual. When the difference in illumination from above is great, Strasburger's own experiments (*l. c.*, p. 33) show that the organisms do react.

On the other hand, Holt & Lee (1901), using a similar prism, found, under similar conditions, that the negative organism, Stentor, does, on the whole, tend to gather at the darker side of the drop. This shows that the difference in illumination between neighboring points in this particular experiment was not below the threshold of stimulation for the organism in question. If, as Holt & Lee suppose, a certain amount of light was reflected from the lighter end of the vessel, then the inclination to go to the darker side would be reinforced by Stentor's tendency to turn when the light falls upon its anterior end (see p. 43). The fact that in Strasburger's experiments the organisms remained scattered throughout the drop seems to indicate that this reflected light played no part in his results.

In another set of experiments Strasburger placed his prism over the swarm-spores in such a way that the light came obliquely from the direction of the thick end of the wedge. If the positive organisms now go toward the thicker end of the wedge, they pass toward the source of light, but into a region of decreased illumination; if they go toward the thin end they pass away from the source of light, but into a region of higher illumination. Which will they choose?

Strasburger found that the positive swarm-spores pass toward the source of light, into the region of less illumination. But is not this exactly what we must expect? His former experiment showed us that under the prism the change from light to darkness was so gradual that it produced no effect on the organisms. Hence the direction from which the rays come is left to produce its effect alone, and it produces the usual effect. The organism reacts in the usual "trial and error" way until the anterior end is directed toward the light; then it moves in that direction. Incidentally it comes into a region of less intensity of light, though the decrease is so slight as to produce no effect on the organism.

Parallel considerations hold for the negative organism. Under similar circumstances, if the variation in illumination is very gradual, it directs its sensitive anterior end away from the source of light (by the method of "trial and error") and swims to the opposite side of the drop, incidentally moving into a region of slightly greater (but "unperceived") intensity of illumination. Under similar conditions, as we have seen in the experiment described on p. 39, if the decrease in illumination is marked, the animal swims back into the shadow, though in so doing it passes toward the source of light.

Thus in Strasburger's experiments with the prism the difference in the intensity of light between neighboring regions has been made so slight that they are unmarked by the organism and have no effect upon it. We need not be surprised, therefore, that it reacts as if these differences did not exist; for the organism they do not exist.

The reaction is in this case just what it would be in a higher organism under similar conditions. Let us suppose that the light stimulates strongly the sensitive anterior end, the eyes, of a higher animal or man; it causes pain in the case of man. There will be a tendency (1) to move into less illuminated regions; (2) to turn the eyes away from the light. Suppose that the man is enclosed in a space into which the sun shines obliquely from above, and that the end from which it shines is a little less illuminated than the opposite end, owing to causes similar to those in Strasburger's experiment on the swarm-spores. Suppose that the man is at the end next the sun. He cannot know that the other end is more illuminated, for the only way this would be possible would be for the greater number of rays of light to meet his eye coming from that direction, while by hypothesis all, or a much larger number, of the rays are coming from the opposite direction. He will, therefore, turn his eyes away from the sun, and if he moves will move toward the end away from the sun. After having traversed some distance he may observe, if he is very discriminating, that he is as a matter of fact getting into a region of somewhat greater illumination, and may perhaps reason that the best thing he can do under the circumstances is to keep his eyes turned away from the source of light and move backward to the less illuminated end. But this involves the capability of making fine distinctions, and a considerable degree of intelligence in deciding what to do under the peculiar circumstances. The experiment with the swarm-spores shows that they are incapable of such fine discrimination, or that they are not sufficiently intelligent to know what to do under the circumstances. They give no indication that they notice the greater illumination after having passed to the end away from the light. Their action may be considered perhaps in a certain sense as a "mistake," but it is a mistake which even the highest organism would make, and which could be corrected only after experience of its results.

The results of our study of the light reaction in ciliates and flagellates lead to conclusions which stand in sharp contrast with certain general conclusions in Radl's recent extensive and interesting paper on Phototropism (Radl, 1903). Radl reaches the somewhat extraordinary conclusion that light orients organisms by exercising an actual mechanical pressure upon them. This pressure necessarily disturbs the equilibrium of the body, which is then compelled to change position until equilibrium is restored; the organism is then oriented. The orientation is a consequence of the interplay of two sets of forces, inner and outer; these cannot be in equilibrium until the body has taken a certain position with reference to the pressure exercised by the light (*l. c.*, pp. 151 ff.) The actual turning which induces orientation must be due to the action of a pair of forces (*l. c.*, p. 148). One of these forces is the pressure produced by the light.

Orientation produced in the manner described in the present paper for the reaction of ciliates and flagellates to light, and in the preceding paper for the reaction to heat, could of course not be brought about in the manner supposed by Radl. One of Radl's chief arguments for his view is that "no observation thus far shows that the final orientation is attained by a trial or after an oscillation, but it takes place automatically"* (*l. c.*, p. 141).

The observations on ciliates and flagellates given in the present paper show conclusively that the orientation in these cases *is* brought about through repeated trials. In the statement quoted above Radl has overlooked certain other cases. Thus Strasburger, as we have seen (p. 59), states that after the direction of the light is changed Hæmatococcus becomes reoriented "nach verschiedenen Schwankungen" (Strasburger, 1878, p. 24). Radl himself refers on a previous page (p. 100) to Strasburger's observation of the oscillating movement of swarm-spores under the influence of a variation in light intensity; Rothert (1901,

*"Keine bisherige Beobachtung zeigt ferner, dass die schliessliche Orientierung etwa durch eine Prüfung oder nach einem Schwanken erzielt würde, sondern sie folgt automatisch."

p. 397) has called particular attention to this as a possible factor in the so-called phototropism. Radl also refers (p. 99) to Exner's view that in Copilia the movements of the eyes are in the nature of a trial ("abtasten") of the surroundings. Radl's statement, quoted above, can then hardly be considered strictly accurate, even leaving out of consideration the results set forth in the present paper. In many organisms, doubtless, the reaction to light is of that direct character assumed to be general by Radl. But it may be strongly doubted whether this is what we may call a primitive condition; in other words, whether it does not involve more complicated internal processes than the reaction by "trial and error." In any case, I am convinced that a similar reaction to light by the method of "trial and error" will be shown to exist in many other organisms; it is demonstrated, for example, in Rotifera, in the paper which follows the present one.

Recourse will doubtless be taken to the usual refuge when a sharp concept has been defined to which the phenomena are not found to correspond; the reactions of the ciliates and flagellates will be simply excluded from the tropisms and the definition of the latter maintained in all its pristine purity. Indeed, it may be questioned whether the reactions of infusoria (and Rotatoria) to light are not excluded from phototropism through the definition given by Radl on p. 140, whatever the method by which they are produced. Radl says "Unter phototropischer Orientierung ist die Fähigkeit der Organismen zu verstehen, eine feste Einstellung der Achsen des gesamten Körpers in dem Lichtfelde einzunehmen." Since the ciliate or flagellate (or rotifer) revolves continually on its long axis, and swerves continually toward a certain side, it can hardly be said that the body axes have a "feste Einstellung" with reference to the light. In an explanatory paragraph Radl says that in orientation "immer geht dann der Lichtstrahl durch die (morphologische) Symmetrieebene des Körpers" (*l. c.*, p. 140). This is certainly not true for the ciliate or flagellate (or rotifer), even leaving out of consideration the fact that in the former two groups the animals are usually unsymmetrical. If it be proposed, then, to exclude the light reactions of ciliates, flagellates, and rotifers from the concept of "Phototropismus," one can only agree that this is necessary, in view of the definitions of that concept.

But what is the value of a definition which excludes some of the chief phenomena on which the concept that we are attempting to define is based? And what is the value of a theory that depends on such a definition and that can only be correct so long as we hold to this definition? The phenomena themselves are, after all, the final reference for testing the correctness of any definition or theory; it is the observed phenomena that we are attempting to formulate and explain.

What we desire in the study of animal behavior is (1) a correct description of what occurs; (2) an understanding of the relation of what occurs here to other phenomena, this constituting their "explanation" so far as an explanation is possible. Whether the phenomena when correctly described and understood are found to fall under someone's definition of a tropism is comparatively unimportant; it is only after such correct description and understanding that final definitions can be made. I question much if there has not been undue haste in framing precise definitions for the phenomena of animal behavior, when we know so little about the phenomena in any thorough way. Radl, I believe, makes a fundamental error in attempting to separate "Phototropismus" rigidly from other reactions to light. Thus, he repeatedly cites Euglena as an example of an organism that shows undoubted phototropism. On page 114 he further cites the motor reaction of Euglena when suddenly shaded * as a reaction that has nothing to do with phototropism. As I have shown above, the two are really closely bound up together; the orientation in the "phototropism" is produced through this motor reaction. When the reactions of organisms to light are known in detail, I believe that many other reactions which Radl (p. 114) attempts to separate sharply from "phototropism" will be found closely connected with the reactions that go under that name. I had occasion to point out, in the paper preceding this, on the reactions to heat, that if everything which the organisms do, except the orientation itself, is left out of consideration, the orientation can be accounted for by any theory desired. A thorough study of precisely this point—the relation of "phototropism" to the phenomena supposedly unconnected with it—would, I believe, have saved Radl from marring his otherwise most excellent and useful contribution to the study of light reactions by the proposal of so fantastic a theory to account for the reactions to light; a theory that fairly produces a shock in the mind of the reader when it is reached, coming as it does after Radl's thorough and valuable objective study of many of the phenomena and his exceedingly sane, if somewhat sharp, criticism of other theories. Definition and precise classification are of course valuable at a certain stage of knowledge, but when carried out without a thorough knowledge of the phenomena dealt with they may be a hindrance rather than a help. The thorough knowledge of the phenomena of animal behavior required for this is far from existing at present.

---

* Radl says when "beleuchtet"; this is evidently a slip of the pen.

THIRD PAPER.

REACTIONS TO STIMULI IN CERTAIN ROTIFERA.

REACTIONS TO STIMULI IN CERTAIN ROTIFERA.

In my series of "Studies on Reactions to Stimuli in Unicellular Organisms" and in the foregoing papers I have set forth the reaction methods of many infusoria to varied stimuli. The result has been to show that the reaction method in these organisms is of a peculiar character, differing radically from that required by prevailing theories of the reactions of lower organisms. The essential nature of these reactions, with their implications as to the character of behavior in the lower organisms, will be discussed in the following papers. Before proceeding to this discussion it is important to determine whether the reaction method in the Infusoria differs radically in character from that of Metazoa. For this purpose it seems well to select a group of Metazoa whose habitat and mode of life are similar to those of the Infusoria. In this way differences due primarily to the different plan of structure of the two sets of organisms may perhaps be brought out without the complications arising from different modes of life.

A group of Metazoa much resembling the Infusoria in their mode of life is found in the Rotatoria. As is well known, the members of these two groups are usually found mingled together. They are of about the same size, and both swim about by means of cilia. So great is the resemblance in general habit and in habitat that they were at first classed together, all being given the name of Infusoria. As we know now, however, they are really widely separated in relationship. While the Infusoria are unicellular, the Rotifera are multicellular organisms of a high degree of complexity, possessing many systems of organs, each composed of many cells. In particular, they have a well-developed nervous system.

A comparison of the behavior of these two groups of organisms should show us, therefore, whether there are types of reaction having a high degree of generality, such as is claimed for the theory of tropisms—types that may give a key to the behavior of groups so widely separated in relationship as the two under consideration, which are representatives of the Protozoa and of Metazoa of a fairly high degree of organization.

In the present paper I can attempt to give an account of the behavior of only a few free-swimming species, and that not in an exhaustive manner. I hope to return to an extensive study of the behavior of this interesting group, so as to develop its implications for the theory of animal behavior in general. In the study here set forth observation was directed primarily to the questions of how certain Rotifera react under the stimulus of the agencies which usually give rise to the so-called tropisms—light, chemicals, heat, electricity, contact, etc.—and to these questions the present account will be devoted.

The species whose reactions were examined belong chiefly to the loricate group of free-swimming Rotifera, and include a number of species of the Rattulidæ, several species of Cathypnadæ, two or three species of Euchlanis, *Plœsoma lenticulare*, *Anurœa cochlearis*, and *Brachionus pala*. These were studied as opportunity offered. In most cases the reactions of any one species were not determined with relation to more than two or three classes of stimuli. The behavior of *Anurœa cochlearis* was examined most fully. This species will be used as a type in describing the reactions. I have already given a brief account of the general reaction type in certain species of the Rattulidæ in my monograph of that group (Jennings, 1903).

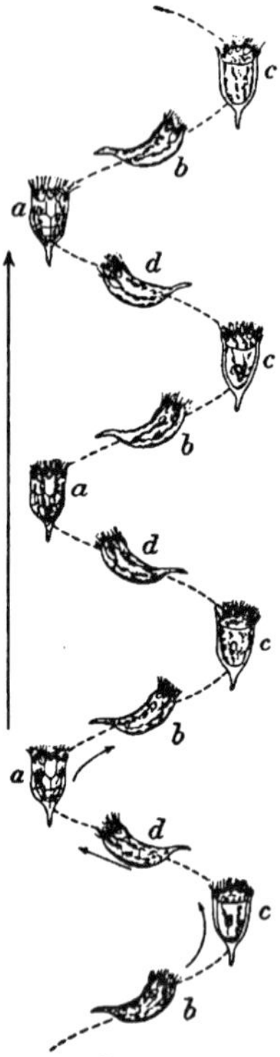

METHOD OF LOCOMOTION.

The free-swimming Rotifera progress through the water in the same manner as the ciliate infusoria. The cilia in the Rotifera are limited to the anterior end, as they are in the peritrichous infusoria. It is interesting to note that the same device is adopted in the one group as in the other, to compensate for irregularities in the form of the body, etc., which might result in swerving from the straight course. This is by revolution on the long axis, causing the path to become a spiral with a straight axis. In the Infusoria the

FIG. 25.*

* FIG. 25.—Spiral path followed in ordinary swimming by *Anurœa cochlearis* Gosse, showing different positions of body in different parts of the course; *a*, dorsal surface; *b*, left side; *c*, ventral surface; *d*, right side. The animal revolves on its long axis over to the right, thus taking successively the positions *a*, *b*, *c*, *d*, *a*, etc. The large arrow indicates the general direction of the course followed; the smaller arrows show direction of progression in certain parts of the course.

organism usually swerves from the straight line toward the aboral side; in the Rotatoria it is usually toward the dorsal side. Well-ordered forward progression would therefore not take place, were it not for the revolution on the long axis, converting the circular course into a spiral one. In the Rotifera the revolution on the long axis is, so far as observed, always over to the right. These relations have been brought out in detail in a previous paper by the present author (Jennings, 1901).

The spiral path thus followed by most of the free-swimming Rotifera may be illustrated in Fig. 25, for *Anuræa cochlearis* Gosse. As will be seen from the figure, the path followed depends upon three factors: (1) the animal continually swerves toward its dorsal side; (2) it progresses; (3) it revolves on its long axis. The result of these three factors is the spiral course. In all these relations the rotifer agrees with the infusorian.

REACTIONS TO STIMULI.

The most general reaction to a stimulus in such a free-swimming rotifer is an accentuation of one of the factors in this course, namely, the swerving toward the dorsal side. The result is to produce a spiral of much greater width than previously existed. This may often be observed when the vessel containing the rotifers is jarred. It is evident that this method of reaction is fitted to enable the rotifer to avoid a small obstacle lying in its path, that is, in the axis of the spiral. When the animal resumes its former method of swimming the axis of the spiral lies in a new direction; the course has thus been slightly changed.

With a stronger stimulus, as when the rotifer strikes against an object lying in its path, the swerving toward the dorsal side may be still more pronounced, while the revolution on the long axis nearly or quite ceases. The result is that the organism swings strongly toward its dorsal side, and when the usual forward swimming is resumed the axis of the spiral lies in a totally new direction (Fig. 26). It thus avoids the obstacle, if the latter is small; if the first reaction does not avoid the obstacle completely the reaction is repeated until the course is sufficiently altered so that the rotifer no longer strikes against the source of stimulus. In some rotifers the increased swerving toward the dorsal side is preceded by swimming backward a short stretch.

In all these points the reaction of the rotifer agrees even to details with that of the ciliate infusorian. There is a difference in the fact that the Infusoria are unsymmetrical and cannot therefore be said to swerve toward the *dorsal* side, as do the prevailingly symmetrical Rotifera. In the Rattulidæ, however, we have asymmetry of a character similar to that found in the Infusoria.

We have dealt thus far specifically only with reactions to simple mechanical stimuli, such as are presented by an obstacle in the path of

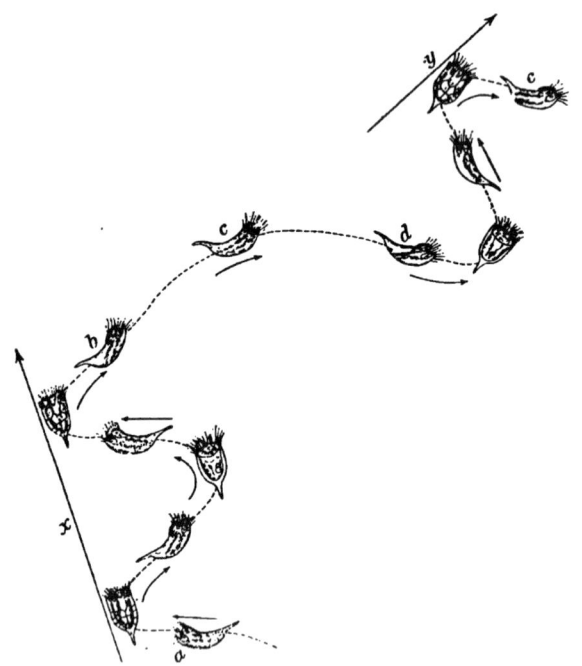

FIG. 26.*

the rotifer or by a simple mechanical jar. This type of reaction under such conditions I have observed for *Diurella tigris* Müller, *D. por-*

---

* FIG. 26 is a diagram of the reaction of the rotifer Anuræa to a strong stimulus, as when it reaches a source of mechanical stimulus or a region where some chemical is dissolved in the water. From *a* to *b* the animal is unstimulated, hence it follows the usual spiral course. At *b* it reaches the stimulating region, whereupon it turns strongly toward the dorsal side, following the arc of a circle, from *b* to *d*. Here it resumes the usual spiral course (*d* to *e*). The large arrow *x* shows the general direction of progression before the stimulus was received; the arrow *y* shows the direction of progression after the reaction has taken place.

callus Gosse, *D. gracilis* Tessin, and a number of other species of Rattulidæ; *Plœsoma lenticulare* Herrick, *Cathypna ungulata* Gosse, *Monostyla bulla* Gosse, *Brachionus pala* Ehr., *Anuræa cochlearis* Gosse, and *A. aculeata* Ehr.

This method by no means exhausts the possibilities of reaction even to a simple mechanical stimulus in these species. They may retract the head and cease swimming, may creep over the surface of the object with which they come in contact, or possibly may sometimes turn otherwise than to the dorsal side when stimulated. Of this latter point I am, however, by no means sure. It is certain that the typical reaction, occurring in the great majority of cases, is that described above.

REACTION TO CHEMICALS.

The reaction given when the organism comes in contact with an area containing a rather strong diffusing chemical was observed in *Metopidia lepadella*, *Anuræa cochlearis*, *A. aculeata*, and *Diurella gracilis*.

The method of experimentation was as follows: A drop of water containing the rotifers was placed on a slide. Near this was placed a drop of N/8 NaCl, and the two drops were connected by a narrow neck. The behavior of the organisms as they came into the region of the neck and thus in contact with the salt solution was observed with the Braus-Drüner microscope. In the species mentioned the reaction was by a sudden turn toward the dorsal side, by which the path of the animal was directed away from the chemical. The reaction is thus of the same character as occurs in the ciliate infusoria.

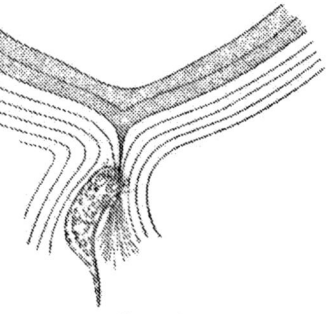

FIG. 27.*

This manner of reaction to chemicals is in both these groups of organisms just what might be expected when the currents caused by the cilia are taken into consideration. In the ciliate, as I have shown

---

*FIG. 27.—Diagram of currents in a nearly quiet Anuræa, showing how a diffusing chemical or an advancing region of warmer water (represented by shading), is drawn out by the ciliary vortex, so as to reach the mouth and the ventral surface before affecting other parts of body.

in a previous paper (Jennings, 1902, *a*), the cilia cause a current coming from the region in front of the organism to pass along the oral surface to the mouth; in this way the oral surface comes in contact with the chemical before any other part is affected. It is not surprising, therefore, that the organism should turn toward the opposite (aboral) side.

In the rotifera the conditions are parallel to those found in the ciliate. The cilia cause a current, which passes to the mouth, on the ventral surface of the body (Fig. 27). The solution thus reaches the ventral surface first, and the reaction is, as might be expected, a turn toward the dorsal side.

It should be distinctly stated that this reaction method is not universal in rotifers even toward chemical stimuli. In some of the larger species, bearing auricles, or with the ciliary apparatus of a very complex character in other respects, varied reactions may occur, which I hope to analyze in another paper.

REACTION TO HEAT.

This was studied in detail only in *Anuræa cochlearis*. A large number of the rotifers were mounted in a shallow trough formed of a slide, as described on p. 12, and one end of the slide was warmed by means of the apparatus shown in Fig. 5. The reactions were then observed with the Braus-Drüner stereoscopic binocular.

As soon as a portion of the slide has been warmed above the optimum, the rotifers in this region turn more strongly than usual toward the dorsal side, so that the course followed becomes a very wide spiral and the animals make little progress. If the heat is increased the revolution on the long axis ceases, while the animals swerve still more strongly toward the dorsal side (Fig. 28), so that they swim in circles, the dorsal surface being directed toward the center of the circle. Usually after circling thus a short time the animals begin again to revolve on the long axis, and dart forward. The direction of this dart forward seems purely random. If it carries the animal out of the heated region the forward movement is continued and the animal escapes. If it does not carry the animal out of the heated region the circling toward the dorsal side is quickly resumed, followed by another dart forward. This is continued either until the rotifer passes out of the heated region or until it is overcome by the heat. Usually, if it does not escape soon from the heated region the circling becomes more rapid and continuous and is kept up till the animal is destroyed by the heat.

If one end of the slide is heated and the animal approaches the heated region from the opposite end the reaction is of the same character as that last described. As soon as the region is reached where the heat acts as an effective stimulus the animal swerves strongly toward the dorsal side, thus beginning to circle, as shown in Fig. 28. If this swerving should continue only till the animal had described a semicircle, then were followed by the forward dart, the animal would of course retrace its original course (or one parallel to it), and would thus escape from the heated region, as happens in the reaction to the electric current (Fig. 29). But the reaction to heat is less precise than this. Usually the animal makes several complete circles before darting forward, and the direction in which it darts seems a random one; sometimes it is toward the heated region, sometimes away from it, sometimes oblique to it. If the path followed leads the animal into the heated region the circling toward the dorsal side, followed by the dart forward, is repeated; while if the path leads away from the heat no farther reaction is caused and the animal escapes. Thus when a large number of the animals swim toward the heated region a considerable number will be seen a little later to swim away again. But in many cases the dart forward carries the animal still farther into the heated region. These specimens then begin to circle again toward the dorsal side, and if the temperature is high they may

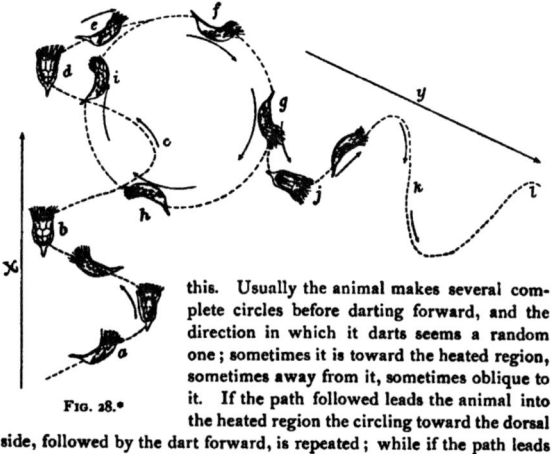

FIG. 28.

* FIG. 28.—Diagram of a reaction to heat in Anuræa. The unstimulated animal at first advances in the general direction shown by the arrow *x*, following thus the course *a* to *e*. The heat is supposed to be advancing from the direction opposite the arrow *x*. When the rotifer reaches the point *e* the heat becomes effective as a stimulus. The animal reacts by turning toward the dorsal side, and continues this so as to describe a complete circle, *f, g, h, i, f*, etc.; often it describes such a circle several times. Finally, at some point in the circular course, as *g*, it resumes the usual spiral course, following thus the path *g, j, l*. Its original course, shown by the arrow *x*, has thus been exchanged for a course having the general direction shown by the arrow *y*.

continue this till death intervenes. In many cases they repeat the dart forward and some escape in this way, while others do not.

The reaction of Anuræa to heat is, therefore, not very precise, and many individuals swim into the heated region and are killed. Those which escape do so through a reaction which is similar to that of those which do not; in the one case the forward movement carries the animal out of the heated region; in the other it does not. The essential point to the reaction is that the animals when stimulated by heat *change their course* (through a " motor reflex"). This changed course naturally is an advantage, and in accordance with the laws of probability carries some of the organisms away from the source of danger. Others, likewise in accordance with the laws of probability, are carried even by the changed course toward the heated region, where they may be killed unless a repetition of the " motor reflex" with its change of course carries them finally away. The reaction is by the method of " trial and error," and is not always successful.

Altogether, the reaction of the rotifer Anuræa to heat is of a character similar in principle to that of Oxytricha (Fig. 7, p. 16). The direction of turning depends on an internal factor; the reaction takes the form of " a motor reflex," and is by no means compatible with the typical tropism schema.

REACTION TO LIGHT.

In light, as I have already set forth in the account of the reactions of Stentor, we have a stimulating agent of a different character from that found in chemicals or in heat, since the distribution of the stimulating agent is not affected by the currents of water produced by the motor organs of the animal. There is thus no reason in the distribution of the stimulating agent to favor a turning toward one side rather than the other.

I have been able to study accurately the light reaction in but one rotifer, *Anuræa cochlearis* Gosse. The conditions necessary for precise observation of the nature of the reaction are very difficult to fulfill, and the usual movements of the animals are such that the nature of the reaction is obscured. As will be recalled, the organism is normally swimming rapidly in a spiral, continually swerving toward its dorsal side. This in itself is very confusing when one attempts to observe just how the organism turns when stimulated. When light is thrown upon it, or when the direction of light falling on it is changed, the response is usually not given at once, and when it does occur, as we shall see, it may be in the form of an accentuation of certain features of the normal movement. From these conditions it results that it is exceedingly difficult to tell, after a reaction to light has clearly occurred, just how the reaction took place. Of course, only sharply defined positive observations are of value in deciding between two opposing possibilities; hence, although I have studied a number of other rotifers in this connection, I give the results only where absolutely sure of them. But in the two or three other rotifers I have examined in this connection the reaction is apparently the same as that in *Anuræa cochlearis*, to be described at once.

The specimens of *Anuræa cochlearis* studied had been in a small aquarium in the laboratory some months, and were distinctly negative to light, gathering at the side of vessel farthest from the window. The freshly collected animals are, I believe, usually positive to light.

These negative individuals were placed in a small flat-bottomed rectangular glass vessel, on a dark background, in a dark room. At opposite sides of the vessel and somewhat above were clamped two incandescent electric lights, *A* and *B*, at a distance of about 10 inches, in the manner described for Stentor (p. 41 and Fig. 15). One of these lights could be extinguished while the other was simultaneously turned on. In this way the direction of the light falling on the rotifers could be reversed.

When only one of the lights, as *A*, was turned on, the Anuræas all collected at the opposite side of the vessel, next to *B*. When *A* was extinguished and *B* turned on, they turned and swam in the opposite direction, toward *A*. By reversing the direction of the light while the animals were crossing the vessel their course could be reversed while in full career.

Focusing the Braus-Drüner on the vessel, and reversing the lights when the animals were well in the field of observation, the following could be observed: Some turned at once, with some sharpness, *toward the dorsal side*, the turning continuing until the direction of swimming was reversed and the animals were again swimming away from the light (Fig. 29). In these cases the direction of turning was clear and could be observed without great difficulty.

Other individuals continued for a short time to swim in the same direction as before, then turned, either sharply, as just described, or more slowly, in the manner to be described.

Where the turning was sharp, as described above, there was no great difficulty in determining with certainty the nature of the reaction. But in many cases the turning took place more slowly, in the following manner: Either as soon as the light was reversed, or very soon after, the width of the spiral in which the animal was swimming became much greater. In other words, the animal swerved more toward the dorsal side and progressed less rapidly than usual. Thus it described rather wide circles, and the swerving toward the dorsal side increased, while progression and revolution on the long axis had largely ceased. After this circling had continued for some time, the swerving toward the dorsal side apparently continually increasing, it was found that the anterior end was directed away from the source of light; *i. e.*, the direction of swimming had been reversed, and the animal was moving away from the light.

It is obviously very difficult to be entirely certain of all that has happened during this period of extensive circling, as a matter of direct observation. But the evidence seems to show clearly that the essential point in changing the course is the swerving toward the dorsal side. The following facts all point to this conclusion : (1) In the individuals which turn at once it is possible to be entirely certain that the turning is toward the dorsal side. (2) In the individuals which are circling it is entirely clear that the swerving toward the dorsal side is greatly increased, and there is no evidence of turning in other directions. The only difficulty is that one cannot follow every evolution and be certain that nothing else has occurred. (3) Analysis of this same reaction when given in response to other stimuli, where the conditions are more favorable for observation, shows that it does consist of an increased swerving toward the dorsal side, with a decrease, or an entire stoppage for a time, of the forward motion. There is, then, no reason to think that the reaction contains other factors when performed under the influence of light. The reaction is indeed clearly the same as that described for Euglena on p. 53, and illustrated in Fig. 21 ; a similar analysis could be given for the reaction of Anuræa.

It may be considered certain, therefore, that in *Anuræa cochlearis* the reaction to light is similar to the reaction to other stimuli, and that the orientation is brought about by a turning toward the dorsal side. The reaction is, therefore, not due to the direct effect of the light on the motor organs ; the direction of turning is determined not by external factors, but by internal factors. The reaction to light in the rotifer, like that in the infusorian, takes place by the method of "trial and error."

REACTION TO THE ELECTRIC CURRENT.

A considerable number of different species of the rotifera were subjected to the continuous electric current without the production of any characteristic reaction. A current was used which could be graded in strength from practically zero to one that was destructive, but no reaction comparable to that found in the ciliate infusoria was produced. On making or breaking the current the animals frequently contracted quickly, and if the current was very strong, the head was completely retracted and the animal sank to the bottom. But there was no orientation and the animals did not swim toward either electrode. These negative results were obtained with several of the Philodinadæ (Rotifer, Philodina), some species of Euchlanis and Salpina, *Noteus quadricornis*, and a number of the Rattulidæ.

In *Hydatina senta* there is a reaction of a peculiar character which perhaps furnishes a clue to the cause of the more pronounced reaction to be described for Anuræa. With a current of moderate strength, such as that to which Paramecia react most markedly, Hydatina shows no reaction except when the head is directed toward the anode. But in this position the animal at once retracts its cilia and sinks to the bottom. Thus a Hydatina may swim freely about in water through which the current is passing, provided it swims toward the cathode, or transversely, or obliquely; as soon, however, as it turns its head toward the anode it stops swimming and sinks to the bottom. Thus if an electric current is passed through a preparation containing a large number of specimens of Hydatina, many will be seen swimming toward the cathode and others at all sorts of angles with the current, but none toward the anode. This is a phenomenon akin to what I have elsewhere called the production of orientation by exclusion. If organisms are prevented from swimming in any direction but one, after a time, provided the course is frequently changed, all that are swimming will be found moving in that one direction. This condition is realized, as I have shown in the first of these contributions, in the reactions of infusoria to heat and cold. But in the reactions of Hydatina to the electric current the "exclusion" is less complete than in the cases just mentioned; the animal may swim in any direction *except* one.

The fact that the head is retracted when directed toward the anode and not in other positions indicates that there is a greater stimulation at the anode than elsewhere. This agrees with much that is seen in the reactions of infusoria to the current. After Hydatina has sunk to the bottom with anterior end to the anode, it repeatedly makes attempts to unfold its cilia. But scarcely have they begun to operate when they are withdrawn again. Each time that they are uncovered for an instant, however, they turn the animal a little toward its dorsal side. Thus, after a considerable number of attempts to unfold the cilia, the head has become turned away from the anode; then the cilia are spread out and the animal goes on its way until it is so incautious as to turn its head again toward the anode.

Anuræa cochlearis shows marked electrotaxis similar to that found in the infusoria. When the continuous current is passed through a preparation containing large numbers of this species, all orient quickly and swim toward the cathode. They thus agree, so far, in their reaction to the electric current, with the ciliate infusoria.

The question as to the mechanism of the electrotactic reaction in the rotifer is of interest when one compares the structure of these animals with that of the ciliate infusoria. The Rotifera, in place of having cilia scattered over the entire body, are furnished only with a group of cilia at the anterior end. In the Ciliata it is usually possible to distinguish functionally two groups of cilia (1) the *adoral* cilia, about the mouth and oral groove, or at the anterior end; (2) the body cilia, scattered over the body. The cilia of the rotifers correspond functionally with the adoral cilia of the Ciliata.

Pearl (1900), Wallengren (1902-1903), and others have shown that in the electrotactic reaction of the ciliates the two sets of cilia are in many cases from a functional standpoint differently affected. The adoral cilia react under the influence of the electric current in such a way as to tend to turn the organism toward the aboral side; that is, they tend to produce the same reaction which the organism gives in response to most other stimuli, a reaction not in harmony with the tropism schema. The body cilia, on the other hand, are differently affected on the different sides or ends of the organism; those on the part of the body directed toward the cathode striking in one direction; those on the part directed toward the anode striking in a different direction. The result is that the organism, through the action of the body cilia, tends to become directly oriented in a way that is in harmony with the tropism schema. (For details, see the papers cited.) In those ciliates in which the body cilia are much reduced, as in the Hypotricha, the turning is determined throughout by the adoral cilia, so that the orientation does not take place in accordance with the tropism schema, while in some others, such as in Paramecium, the influence of the body cilia is predominant, and the turning is in accord with the theory of tropisms.

What conditions shall we find in the Rotifera, where the only existing cilia seem to agree functionally with the adoral cilia of the Ciliata?

As we have seen, Anuræa swims as a rule in rather wide spirals, swerving strongly toward the dorsal side and revolving on its long axis (Fig. 25). When the electric current suddenly acts upon it the organism at once turns strongly toward the dorsal side, continuing the turn until its head is brought toward the cathode, toward which it swims (Fig. 29). In some cases, as we shall see later, several reactions are necessary for bringing the body in line with the current, but these are as a rule very quickly accomplished.

If, while the animals are swimming toward the cathode, the current is suddenly reversed, the animals again turn strongly toward the dorsal side, continuing the turning until their position is reversed and the heads point toward the new cathode (Fig. 29). In many cases the turning is continued still farther, so that the head of the animal describes a complete circle; indeed, this may continue so that the animal whirls around several times, always towards the dorsal side. The reaction thus far is the same as that produced by heat (Fig. 28). In reacting to the electric current the whirling finally ceases with ante-

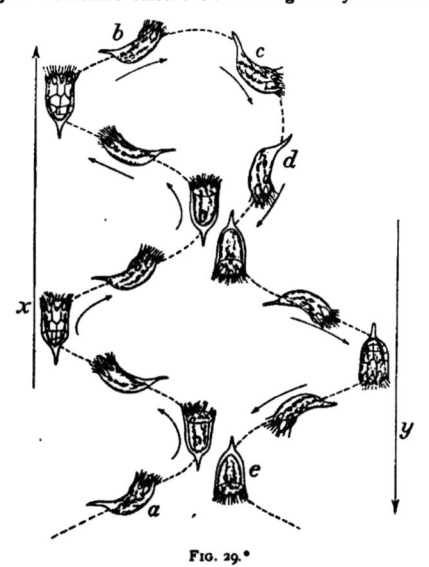

FIG. 29.*

rior end directed toward the new cathode. The animal then swims forward in the direction so indicated. These turnings, even when several times repeated, require but a moment, so that very soon practically all the specimens are swimming toward the new cathode. The

* FIG. 29.—Diagram of method by which Anuræa becomes oriented to rays of light, or to the electric current. Taking the latter for example, the animal is at first swimming toward the cathode, in direction indicated by arrow x; it thus follows a spiral path from a to b. At b the electric current is reversed. The animal thereupon swerves strongly toward its dorsal side, describing a semicircle, b, c, d, until its anterior end is directed toward the new cathode, in the opposite direction from before. It now follows the spiral path d to e, in the general direction indicated by the arrow y. The facts are similar for the reversal of light, or for the reaction when the current or the light is first set in operation.

reaction in Anuræa is in a general view as striking and clear-cut as that of Paramecium.

Thus in the rotifer Anuræa the orientation to the continuous electric current is produced through a motor reaction, the essential features of which are determined by the structure of the organism. The organism turns always toward the dorsal side, continuing or repeating the turning until the anterior end is directed toward the cathode. In these respects it agrees with hypotrichous Ciliata, where the direction of turning is determined by the action of the adoral cilia. The method of reaction is quite incompatible with the tropism schema.

SUMMARY.

The reactions of those Rotifera of which an account is given in this paper take place in a manner essentially similar to the reactions of the ciliate infusoria.

In the reactions to mechanical stimuli, to chemicals, and to heat, orientation is not a striking feature. The organism turns when stimulated toward a structurally defined side—as a rule toward the dorsal side; in this way it avoids the source of stimulus.

In the negative reaction to light the organism becomes oriented with anterior end directed away from the source of strongest light, but this orientation is brought about in the same manner as in Stentor; the animal turns toward the dorsal side without relation to the side on which the light strikes it, and continues the turning or repeats it until the anterior end is directed away from the source of light.

To the continuous electric current the rotifer Anuræa orients and swims directly toward the cathode. The reaction is brought about in the same manner as the orientation to light. When the current is made or reversed the animal turns toward the dorsal side and continues the turning until the anterior end is directed toward the cathode.

Thus the direction of turning is throughout dependent on an internal factor, not primarily on the way in which the stimulus impinges on the organism. These reactions of the Rotifera are thus inconsistent with a theory of tropisms which regards orientation as a primary feature of the reactions, and which holds that the action of the stimulating agent is a direct one on the motor organs of that part of the body on which it impinges. The reactions of the Rotifera, so far as described in the preceding pages, are brought about, like those of the infusoria, by what may be called the method of "trial and error." The reaction to any stimulus is of such a nature as to head the organism successively in many different directions. That direction is followed in which there is no stimulus to induce further turning.

FOURTH PAPER.

THE THEORY OF TROPISMS.

CONTENTS.

	PAGE.
To what Extent does the Theory of Tropisms throw Light on the Behavior of Lower Organisms?	91
Essential Points in the Theory of Tropisms,	92
Reactions to Mechanical Stimuli,	94
Reactions to Chemicals,	96
Reactions to Heat and Cold,	98
Reactions to Changes in Osmotic Pressure,	98
Reactions to Light,	98
Reactions to Gravity,	100
Reactions to Electricity,	100
Résumé and Discussion,	103
Summary,	106

THE THEORY OF TROPISMS.

TO WHAT EXTENT DOES THE THEORY OF TROPISMS THROW LIGHT ON THE BEHAVIOR OF LOWER ORGANISMS?

The writer has been engaged for a number of years in a study, as exact and detailed as possible, of the behavior and reactions of a number of lower organisms. While the results obtained have not, as a rule, agreed with the view that the behavior of these organisms is determined largely in accordance with the prevailing theory of tropisms or taxis, he has not discussed their relation to this theory in detail. This was because of the possibility that the reactions which he had studied were exceptional, and that further investigation might show after all that the behavior of the lower organisms is largely in accordance with the tropism schema.

At the present time the writer feels that the work which he has done, or which has been done by those associated with him, is of sufficient extent to justify the pointing out of certain general relations. The reactions of ciliate infusoria, which have long been used as the types of illustration for the tropisms, have been examined in much detail, and less extensive studies have been made on the Bacteria (Jennings & Crosby, 1901), the Flagellata, and the Rotifera. The reactions of a flatworm have been studied in much detail (Pearl, 1903), and researches are nearly ready for publication, by investigators associated with the author, on the behavior of Hydra and of the leech, and still other studies are under way. Thorough studies, directed to the observation of the exact movements of organisms under stimuli, have recently been given us by other observers also. It seems, therefore, worth while to bring out, in a preliminary way at least, the relation of the observations made to the prevailing theories of animal behavior. In the present paper this will be limited to a consideration of the theory of tropisms, since this is the theory most widely held.

The great apparent value of the theory of tropisms or taxis lies in the fact that it seems to reduce to very simple factors a large number of the most striking activities of organisms, namely, those involved in going toward or away from sources of stimuli of almost any character. It is a schema, in accordance with which almost any movements of the organism (not purely random) might be supposed to take place.

ESSENTIAL POINTS IN THE THEORY OF TROPISMS.

The two essential features of the theory of tropisms are apparently the following: (1) The movements of organisms toward certain regions and their avoidance of others are due to *orientation; i. e.*, to a certain position which the organism is forced by the external stimulus to take, and which leads the organism toward (or away from) the source of stimulus, without any will or desire of the organism, if we may so express it, to approach or avoid this region. (2) The external agent by which the movement is controlled produces its characteristic effect directly on that part of the body upon which it impinges. It thus brings about direct changes in the state of contraction of the motor organs of that part of the body affected as compared with the remainder of the body, and to these direct changes are due the changes shown in the movements of the organism. This is brought out clearly in the quotation from Verworn given on page 8. Loeb (1900, p. 7) sums up the theory of tropisms as follows:

> The explanation of them [the tropisms] depends first upon the specific irritability of certain elements of the body surface, and, second, upon the relations of symmetry of the body. Symmetrical elements at the surface of the body have the same irritability; unsymmetrical elements have a different irritability. Those nearer the oral pole possess an irritability greater than that of those near the aboral pole. These circumstances force an animal to orient itself toward a source of stimulus in such a way that symmetrical points on the surface of the body are stimulated equally. In this way the animals are led without will of their own either toward the source of stimulus or away from it.

Holt & Lee (1901) again bring out our second point, as applied to reactions to light, with especial clearness:

> The phenomena that have led to such an assumption can be satisfactorily explained on the simpler theory that every ray of light impinging on an organism *stimulates at the point on which it falls*,[*] and in proportion to its intensity. * * * The light operates, naturally, on the part of the animal which it reaches. The intensity of the light determines the sense of the response, whether contractile or expansive, and the place of the response, the part of the body stimulated, determines the ultimate orientation of the animal." (Holt & Lee, 1901, pp. 479-480.)

The theory of tropisms as above set forth depends upon the reflex contractility of the motor organs when affected by certain stimuli. An attempt has been made to give it a still simpler form in a recent paper by Ostwald (1903). Ostwald would omit even the factor of reflex irritability, holding that the turning which brings about orientation is a mechanical result of differences in the internal friction of the water or

---
[*] Original not italicized.

similar physical differences. The organism is considered to continue to move its motor organs in exactly the same way after the external change (usually called a stimulus) has taken place; the reason for turning lies only in the different mechanical effect produced when the motor organs act on a medium of greater or less internal friction than before.

It is difficult to conceive how anyone having any acquaintance with the movements of organisms could propose such a theory as that of Ostwald, and indeed this author states (p. 24) that his account is purely theoretical, and that he has not attempted to test his theory by experiment. We need not, therefore, dwell upon the theory, further than to point out the fact that the reactions of many of these lower organisms have been studied thoroughly, and the reflex movements which they perform when subjected to directive stimuli have been fully described, and that these movements are entirely incompatible with such a theory as that which Ostwald sets forth.* If details are desired, it may be pointed out that all the observations brought in the following that are inconsistent with the theory of tropisms as dependent upon direct stimulation of the motor organs are *a fortiori* inconsistent with such a theory as that of Ostwald.

We may, then, turn to the theory of tropisms as set forth in the above quotations from Verworn, Loeb, and Holt & Lee. Diagrams illustrating the method of action of a stimulus, on this theory, are given in the first of these contributions (Figs. 1 and 2).

How far does this theory go in explaining the behavior of the lower organisms? "Tropisms" has become the keyword everywhere in animal behavior; it is supposed to furnish a ready explanation of most of the puzzles which we here encounter. How far is this justified?

This question can be answered only by accurate observation of just what organisms do under the influence of stimuli. The theory of

---

*Some of the assumptions which Ostwald makes as a basis for his physical analysis of the swimming of the lower organisms are so extraordinary as to deserve mention as curiosities. He states, for example, that as a rule the lower swimming organisms which exhibit the tropisms show active movement vertically only upward; he thinks it probable that cases where they have been described as swimming actively downward are errors; that such downward movement is really only passive falling. Yet everyone who has worked with Paramecium or other Ciliata must know how far from the facts is this idea. In a vertical tube Paramecia hasten as freely, and almost as frequently, downward as upward. These infusoria by no means collect at the top in a vertical tube so regularly as the literature on geotropism might lead one to suppose; Paramecia of this region at least are almost as likely to collect at the bottom as at the top. And there is little more difficulty in Paramecium in distinguishing an active movement downward from a passive one than there is in man. From my own observations I know that parallel statements could be made for many other free-swimming organisms, including Metazoa (Rotifera and Crustacea), as well as Protozoa.

tropisms says that certain definite things happen in the change of position undergone by organisms under the influence of stimuli; that the organisms perform certain acts in certain ways. The problem for the investigator is, then, *Do* these things happen? *Does* the organism perform these acts, in these particular ways? These questions are not metaphysical; they can be answered by observation.

We have now before us a considerable body of exact observations which permit us to answer these questions for a certain number of organisms. We will here attempt to summarize these observations so far as they bear upon the essential points in the theory of tropisms. In particular, we will ask, (1) Is the observed behavior brought about through orientation, in the way the theory of tropisms demands? (2) Does the evidence show that the action of a stimulus is directly upon the motor organs of that part of the body on which the stimulus impinges?

REACTIONS TO MECHANICAL STIMULI.

The reactions to simple mechanical stimuli, as when the organism is touched or struck by a hard object over a certain definite area of the body, of course do not as a rule present the conditions required for the production of a tropism, including a definite orientation. Yet it is important to bring out certain general relations shown in these reactions, as they throw light on the reactions to stimuli of a different character.

Most animals show in one way or another a tendency to avoid sources of mechanical shock. In the higher organisms the reaction usually takes the form of a turning away from the side stimulated. The point which needs to be brought out here is that in ciliate infusoria the direction of turning depends, not upon the part of the body stimulated, but upon an internal factor. Stylonychia turns to the right, whether stimulated on the right side, on the left side, on the dorsal surface, on the anterior end, or by a general unlocalized mechanical shock; and parallel statements can be made for other infusoria. (For details see Jennings, 1900.) We have proof, therefore, that *the action of the stimulus is on the organism as a whole, not merely upon the motor organs of that region of the body stimulated.* Further, it is clear that the response is a reaction of the organism as a whole, not one brought about as an indirect result of the fact that certain motor organs have received a stimulus to contraction.* In these respects, therefore, the reactions to mechanical stimuli are different in character from those assumed to take place in the tropisms, and even in these unicellular organisms the processes taking place must be more complex than the

---
* This fact becomes still more striking when we recall that the reaction takes place in the same way in pieces from any part of the body, from which any given motor organs may have been removed. (Details in Jennings & Jamieson, 1902.)

theory of tropisms assumes. Certainly a reaction of the organism as a unit, in response to a localized stimulus, is a phenomenon of a higher and more complex order than would be a simple contraction or other direct change in the motor organs at the point stimulated.

In the higher Metazoa the reaction to a slight mechanical stimulus at one side is usually a turning either toward or away from the source of stimulus. So long as we do not analyze the process further, this result might be interpreted either as due to the direct response, by contraction, of the muscles primarily affected (thus in accordance with the tropism theory), or as a response of the organism as a whole, dependent, perhaps, on an alteration in its physiological condition brought about by the stimulus. The former interpretation is doubtless much the simpler. But we find in the unicellular organisms that this first interpretation is impossible, and that we are forced to the less simple and definite conclusion that the organism reacts as a whole. Does it not then become probable that in the higher animals the very simple, almost mechanical, explanation is likewise incorrect; that we have in them a phenomenon at least as complex as that found in the unicellular animals? In other words, should we conclude that the reactions in the higher Metazoa are simpler and less unified than in the Protozoa?

Fortunately, however, we are not forced to base our conclusions on general considerations. These reactions have been minutely studied in very few of the bilateral Metazoa, but Pearl (1903) has given us a thorough analysis of the reactions of a flatworm (Planaria). This cannot be taken up in detail here, but we may quote Pearl's conclusion in regard to the positive reaction. This consists in a turning toward the point stimulated, on a superficial view a very simple reaction, one especially well fitted for explanation on the theory of direct action of the agent on the motor organs of the region stimulated. Pearl concludes, after exhaustive study, that the processes in the reaction are as follows:

A light stimulus, when the organism is in a certain definite tonic condition, sets off a reaction involving (1) an equal bilateral contraction of the circular musculature, producing the extension of the body; (2) a contraction of the longitudinal musculature of the side stimulated, producing the turning toward the stimulus (this is the definitive part of the reaction); and (3) contraction of the dorsal longitudinal musculature, producing the raising of the anterior end. In this reaction the sides do not act independently, but there is a delicately balanced and finely coordinated reaction of the organism as a whole, depending for its existence on an entirely normal physiological condition. (*l. c.*, p. 619.)

Further studies carried on under the direction of the writer, and soon to be published, will show that in certain other bilateral Metazoa it is equally impossible to explain the simple turning toward a stimulus as a direct reaction of the motor organs of the part stimulated.

It will be important to keep in mind the nature of the reactions to mechanical stimuli, especially in the infusoria, in considering the reactions which are more usually classed among the tropisms.

REACTIONS TO CHEMICALS.*

The reactions to chemicals have been studied by the present author and those associated with him in many Ciliata, in certain Flagellata, in the Bacteria, the Rotifera, and the flatworm; further studies, not yet published, have been made on other organisms. Now, in regard to our first question, as to orientation, the following must be said: In no case has the typical reaction been found to take the form of an orientation, such as is demanded by the theory of tropisms. In the ciliates, flagellates, and rotifers the reaction has been found to take the form of a "motor reflex," a backing followed by a turning toward a certain structurally defined side, without regard to the direction from which the chemical is diffusing. It is this motor reflex that causes the organisms to collect in the region of certain chemicals, and to avoid others. (Details in Jennings, "Studies," Nos. I-X.)

In the Bacteria the results of our work (Jennings & Crosby, 1901) are in agreement with those of Rothert (1901). Here, again, in the gatherings in certain chemical solutions, or in the avoidance of others, there is nothing resembling an orientation in the lines of diffusion. The phenomena are brought about through a reaction of the same essential character as the motor reflex of the infusoria, but still simpler. The essential point is that the Bacteria, when stimulated chemically, reverse the direction of movement. (Details in the papers just cited.)

In the flatworm the results of the thorough study of the chemical reaction by Pearl (1903) may be given in that author's own words:

Planaria does not orient itself to a diffusing chemical in such a way that the longitudinal axis of the body is parallel to the lines of diffusing ions. Its reactions to chemicals are motor reflexes identical with those to mechanical stimuli. The positive reaction is an orienting reaction in the sense that it directs the anterior end of the body toward the source of stimulus with considerable precision, but it does not bring about an orientation of the sort defined above. (Pearl *loc. cit.*, p. 701.)

For details, the original paper of the author quoted must be consulted. It may be added that this positive reaction, by which the anterior end is directed toward the source of stimulus, is identical with that which takes place in response to a single mechanical stimulus. This is analyzed above (p. 95).

Are there any precise and detailed observations which support the idea that the reaction to chemicals is ever a typical tropism? Before

---

*For a statement of the theory of tropisms as applied to chemicals, see Loeb (1897, p. 442) and Garrey (1900, pp. 292, 293).

the method of reaction by a "motor reflex" had been described the reactions to chemicals had been referred in a general way to the tropism schema, but critical observations, which would differentiate between the possibilities, have been lacking. It is necessary to use the greatest caution in this matter, as is shown by the case of Chilomonas. Garrey (1900), although he stated that "a study of the mechanics by which the organism is oriented or by which it is prevented from moving from the ring into the stronger acid of the clear area, or the weaker acid surrounding the ring, proved fruitless," nevertheless concluded that the reaction in this animal was a case of typical tropism. In a paper published in the same number of the same journal (Jennings, 1900, a), I showed that when the mechanism of the reaction *is* worked out, this conclusion does not hold, but that the reaction takes place through a motor reflex, similar to that in the Ciliata. In cases, therefore, where the mechanism of the reaction (that is, the exact movements which the organism performs) has not been worked out, conclusions as to the nature of the reaction are of little value. The only case of which I know in which an author acquainted with the method of response by a "motor reflex" maintains, on the basis of observation, a reaction of unicellular organisms to chemicals in accordance with the theory of chemotropism, is the case of Saprolegnia swarm-spores, as described by Rothert (1901). In this case we are dealing with very minute organisms, and Rothert has made no attempt to give an analysis of the relation of the direction of turning to the differentiations in the body of the organism, such as we found to be necessary above for Chilomonas before the real nature of the reaction could be determined.

Thus it is clear that cases of true chemotropism, in accordance with the general tropism schema, are exceedingly rare, if they exist at all. In almost all the lower organisms in which this matter has been carefully studied it has been demonstrated that the reaction to chemicals is of a different type from that demanded by the tropism theory.

In the discussion so far we have devoted attention particularly to the question of orientation. When we examine the second question proposed, as to whether the stimulus acts directly upon the motor organs of that part of the body on which it impinges, we find the answer somewhat less clear than it was in the case of mechanical stimuli. It is true that in the Infusoria and Rotifera the direction of turning is, as in the case of mechanical stimuli, always toward a structurally defined side, without regard to the direction from which the chemical is diffusing, so that at first view it seems beyond question that the reaction is *not* due to the direct action of the stimulus on the motor organs of the region on which it impinges. While this conclusion is highly probable, the observed facts do not demonstrate it for chemical stimuli as they do for mechanical stimuli. This is owing to the fact that the organism determines for itself the region in which it shall be stimulated by a chemical in solution, as well as the side toward which it shall turn. Now, it appears that the side on which, by its own activities, it is, as a rule, first stimulated by a chemical, is (usually, at least) opposite that toward which it turns. (For details, see Jennings, 1902, *a*.) It could be contended, therefore, that the direction of turning, in the case of chemical stimuli, is a result of the direct action of the stimulating agent on the side stimulated. Such a contention would have little general significance, however, in view of the fact that the same reaction occurs as a response to various other stimuli, where this explanation is quite impossible.

In certain higher organisms, researches which were made under the direction of the writer and are soon to appear will show (1) that chemical stimuli may produce local contractions in the part of the body with which the chemical comes in contact; (2) that these local contractions have little to do with the characteristic behavior of the animals when subjected to chemicals.

REACTIONS TO HEAT AND COLD.

Reactions to heat and cold have been fully discussed in the first of these contributions. It is only necessary, therefore, to point out that the results are in almost every detail parallel with those for reactions to chemicals, and in the same way and to the same degree inconsistent with the theory of tropisms. In organisms higher than the Infusoria and Rotifera, the reactions to heat and cold have been very little studied from the present point of view.

REACTIONS TO CHANGES IN OSMOTIC PRESSURE.*

In the ciliate infusoria the reactions to differences in osmotic pressure are identical with those to chemicals, save that the organisms are much less sensitive to osmotic changes. (Details in Jennings, 1897 and 1899.) The bearing of these reactions on the theory of tropisms is, therefore, the same as was brought out above in the discussion of the reactions to chemicals.

REACTIONS TO LIGHT.

The phenomena shown in the reactions of organisms to light have perhaps formed the chief basis for the theory of tropisms. There is usually a definite orientation shown by the organisms; they move with the axis of the body parallel with the light rays either to or from the source of light. The existence of such orientation forms the basis of the theory of tropisms, and has been considered sufficient in itself as a proof of the

---

* "Tonotaxis," Massart; "Osmotaxis," Rothert.

theory. Yet the theory makes certain definite statements as to the cause of the orientation and the way in which it is brought about. These statements are open to observation and experiment. In most bilateral animals it is indeed difficult to really test the theory. This is because these animals may turn directly toward either side under the influence of light, and it is difficult to tell whether this turning is due to the direct action of the light on the motor organs or to a reaction of the organisms as a whole induced by some change in physiological condition brought about by the light. But in the ciliate infusoria we find a set of organisms so constituted as to permit us to bring the theory to a direct test. These organisms are unsymmetrical, and, as we have seen, the usual reaction is by a motor reflex involving a turning toward a structurally defined side. We can, therefore, arrange our experiments in such a way that the turning demanded by the theory of tropisms shall be the opposite of that usually produced in the reaction of the organism as a whole, and observe the results.

This is what was done with *Stentor cæruleus*, as described in the second of these contributions. The result, as we have seen, is that the organism turns toward a structurally defined side, without regard to what is demanded by the theory of tropisms. The same result was obtained with a number of flagellates and with a bilateral Metazoan— the rotifer *Anuræa cochlearis*.

Thus, in these cases, it is impossible to interpret the reactions as due to the direct action of the light on the motor organs of the side on which the light impinges. The response is as clearly a reaction of the organism as a whole as is the reaction to mechanical stimuli.

Now that it has been shown that orientation to light does occur in some cases in a manner quite at variance with the postulates of the theory of tropisms, and this in organisms widely separated in structure and classification, it can no longer be held that orientation is, *per se*, a proof of the tropism theory. In other words, cases in which orientation takes place, but in which the manner in which it is brought about has not been observed, can not be assumed as cases of typical tropism, due to the direct action of the light on the motor organs of the side affected. The reactions of flagellates and swarm-spores to light, as described by Strasburger (1878), have long been considered types for the tropisms. In the second of these contributions I have shown that in Euglena and Cryptomonas (the latter being one of the organisms studied by Strasburger) the reactions do not take place in accordance with the tropism schema. So far as can be judged from Strasburger's account the reactions of the swarm-spores take place in essentially the same manner as in the flagellates. As Rothert (1901) has pointed out, there are many details in Strasburger's account which suggest that the explanation on the tropism schema is incorrect. These details become intelligible as soon as we understand the real method of reaction as set forth in the second of these contributions. The assumption that the reaction is a typical tropism, when only the fact of orientation is known, is as likely to fall to the ground in other cases as in those just mentioned.

The reaction of bacteria to light, as shown by *Bacterium photometricum*, described by Engelmann (1882), is a typical example of a reaction through a motor reflex not fitting the tropism schema at all.

To sum up, it is clearly shown in certain cases that the reaction to light takes place in a way that is not consistent with the theory of tropisms, and this is true in some cases where a pronounced orientation exists. In many cases of orientation, where it is supposed that the theory of tropisms holds, this is an assumption, for the observations which would decide the matter are lacking.

REACTIONS TO GRAVITY.

In no case have the exact movements of unicellular organisms in response to gravity been worked out in the manner in which this has been done for the reactions to mechanical stimuli, chemicals, heat, light, and electricity. We are, therefore, without the requisite data for deciding whether these reactions agree with the theory of tropisms or do not.[*]

In the higher organisms in which the positive and negative reactions to gravity have been observed (starfish, holothurians, flies, insect larvæ, etc.), the conditions are so complex that, so far as I am able to see, observations which are crucial for deciding as to the mechanism of the reactions have not been made and perhaps can not be made.

REACTIONS TO ELECTRICITY.

As we have seen in the third section of this paper, the reactions of the rotifer to the continuous electric current do not take place in accordance with the theory of tropisms. Anuræa shows a striking orientation to the electric current, swimming directly to the cathode. Yet this orientation is brought about in a way that is quite inconsistent with the tropism schema. The reaction takes place through a "motor reflex," the direction of turning is determined by an internal factor, and not by the way in which the current strikes the organism. The reaction can only be interpreted, therefore, as a reaction of the organism as a whole.

---

[*] In a forthcoming paper by the author, based on work done since the above was written, it will be shown that the reactions of Paramecium to gravity take place in the same way as the reactions to most other stimuli, so that they do not agree with the theory of tropisms.

In the reactions of the ciliate infusoria to the constant electric current, however, we have, if nowhere else, phenomena which show to a certain extent clear-cut and undoubted agreement with the theory of tropisms. To this agreement with the theory of tropisms much of the widespread adherence to the tropism theory for reactions in general is doubtless due. The reaction of infusoria to the electric current is considered a type for the other reactions of organisms.

Yet, in deciding to what extent the theory of tropisms helps us to understand the behavior of organisms, certain striking facts in regard to the reaction to the electric current need to be taken into consideration. These are the following:

(1) The reaction to the electric current never takes place in nature. As has been repeatedly pointed out, the electric reaction is a product of the laboratory; it is a reaction which the organism never gives under normal conditions. This being true, it should not be made the type for reactions in general unless it can be shown clearly that the characteristic features in the effects of electricity on organisms are present also in the effects of other agents. Otherwise we may fall into the same error that would exist if we considered the contortions of a person who had grasped the electrodes of a powerful battery as a type of human behavior in general.

(2) But examination shows that the characteristic features of the effect of electricity on organisms are not present in the case of other stimuli. The electric current, as the experiments of Kühne (1864) and Roux (1891) have shown, polarizes the cell. That is, it divides it into halves, differing in chemical reaction. One half, in the case described by Kühne, had apparently an acid reaction, the other half an alkaline reaction. In its effects on free-swimming organisms a similar polarity is shown. In Paramecium, for example, the cilia on one half of the body (where the current is entering) are caused to take a certain position, while those on the other half (where the current is leaving) take the opposite position. *No other agent known produces these polar effects*, either chemically or in the effect on the motor organs. Yet it is to exactly these effects that the orientation which makes this reaction the type for the tropism theory is due.

If other agents produce these effects why are they not known and described? There is no great difficulty in observing these effects with the use of the electric current. Just as exact studies have been made of the reactions to other stimuli; yet, so far as I am aware, no one has ever described any other stimulus as giving these characteristic polar effects. On the contrary, the reactions to other stimuli are well known *not* to show these characteristic phenomena.

Since, therefore, the characteristic phenomena of the reaction to the electric current are not found in the reactions to other stimuli, it seems a perversion to make the electric reaction a type for all others. The reaction of the infusoria to the electric current takes in its characteristic features a unique position among the reactions of the organism, requiring special explanation.

(3) In the response of the infusoria to the electric current there appears also the same type of reaction that occurs as a response to other stimuli, but obscured by the phenomena peculiar to the effects of the current.

This fact, that the reaction to the electric current is of a dual character, that the peculiar effects of the current are, as it were, superposed upon the usual method of reaction, is not usually so clearly recognized as it deserves to be.

If the constant current is passed through a preparation containing large numbers of some species of the Hypotricha, as Stylonychia or Oxytricha, it will be found that the animals, practically without exception, attain their orientation by turning toward the right side, thus reacting as they would to any other stimulus. Further, if after they are oriented the direction of the current is reversed, the animals all, without exception, attain their new orientation (with anterior ends in the opposite direction) by whirling toward their right sides. Thus, so far, the reaction to the electric current is identical with that to other stimuli, and the direction of turning is determined by an internal factor, not by the way in which the current strikes the organism. In these respects the Hypotricha agree with the Rotifera.

But exact observation shows that in the Hypotricha there is another factor involved in the reaction. The characteristic polarizing effect of the current appears in its action on the motor organs that are distributed over the body surface; those on one half of the body strike in one direction, those on the other half in the opposite direction. Part of these motor organs, therefore, assist in turning the organism in its usual way (to the right); part oppose this turning. The result is that in certain positions the turning to the right is opposed by the stroke of a large number of cilia, so that the turning takes place more slowly than usual. Nevertheless, in the Hypotricha, the determining factor in the reaction to the electric current is almost throughout the same as in the reaction to other stimuli; the direction of turning is determined by internal factors, as a reaction of the whole organism, not by the direction in which the current strikes or passes through the organism. (Details in the work of Pearl, 1900.)

If in place of one of the Hypotricha we experiment with an infusorian in which the cilia cover closely the whole surface of the body, as Paramecium, the peculiar polarizing effect of the current on the cilia of the two halves of the body becomes much more powerful, because the number of cilia affected in this way is much greater. The result is that it almost alone determines the nature of the reaction. The direction of turning is, therefore, determined by the way in which the current strikes the body, as required by the theory of tropisms. But it should be recognized that this is by no means universal among the infusoria; in doubtless fully as many cases the direction of turning is determined, even under the electric stimulus, by an internal factor.

This peculiar dual character of the reaction to the electric current—one strong factor being due to the inherent tendency of the organism to turn in a certain definite way, without regard to the way in which the stimulating agent impinges upon it—has been studied in detail in recent contributious by Pearl (1900), Pütter (1900), and Wallengren (1902 and 1903). We may perhaps compare it, without indicating any similarity in details, to the behavior of a person who has taken hold of the electrodes from a powerful induction coil. He attempts in various ways to free himself from the electrodes. This may be compared with the attempt of the infusorian to perform its usual reaction to strong stimuli. He also undergoes involuntary contortions, due to the action of the electricity on his muscles; these may be compared with the peculiar effect of the electric current on the cilia of the infusoria, causing them to strike in opposite directions on the two halves of the body.

Putting all together, we are not justified in taking the reaction of the infusoria to the electric current as a general type for the reactions of the lower organisms, because in its characteristic features it differs from all the other known reactions. Yet it is exactly these unique features that bring it into (partial) agreement with the tropism schema.

RÉSUMÉ AND DISCUSSION.

We have thus passed in review the reactions of a large number of lower organisms to the commoner stimuli, so far as they are known from exact observations. We have found that as a rule they do not fit into the tropism schema. In the reactions to mechanical stimuli, to heat and cold, to chemicals, to changes in osmotic pressure, orientation is not a primary or striking factor of the reaction; when a common orientation of a large number of organisms occurs, it is a secondary result, due to the fact that the organisms are prevented from swimming in any other direction. In the reaction to light orientation is a striking feature; but the orientation, in certain precisely investigated cases at least, is brought about in a manner which is inconsistent with the tropism schema. In the reaction to gravity the precise reaction method has not been worked out. Only in the reaction to the constant electric current do we have in some organisms a partial agreement in principle with the requirements of the tropism theory, and this agreement is due to an effect on the organism in the production of which the electric stimulus is unique, so far as known. In none of the reactions which have been thoroughly worked out, except partially in the reaction to the electric current, are the phenomena to be explained on the view that the result is due to the direct action of the stimulating agent on the motor organs of the part of the body on which it impinges. In the reactions to mechanical stimuli and to light, and in the reactions to the electric current in some animals, this view is absolutely disproved. The direction in which the organism turns is, in all the well known reactions of unicellular organisms and rotifers (except in a portion of the reactions to the electric current), determined by an internal factor, and predictable from the structure of the organism without any knowledge of the direction from which the stimulating agent is to come.

We should perhaps consider here a modification of the original form of the tropism theory that has been proposed by some authors. This is in regard to the assumption that the stimulating agent acts directly on the motor organs upon which it impinges. For this it is sometimes proposed to substitute the view that the action of the stimulating agent is directly on the sense organs of the side on which the stimulus impinges, and only indirectly on the motor organs through their nervous connection with the sense organs. When thus modified the theory, of course, loses its simplicity and its direct explaining power, which made it so attractive. In order to retain any of its value for explaining the movements of organisms, it would have to hold at least that the connections between the sense organs and motor organs are of a perfectly definite character, so that when a certain sense organ is stimulated a certain motor organ moves in a certain way. When we find, as we do in the flatworm (see the following paper), that to the same stimulus on the same part of the body, under the same external conditions, the animal sometimes reacts in one way, sometimes in another, the tropism theory, of course, fails to supply a determining factor for the behavior.

But can we explain the reaction methods of the infusoria and other animals which, as set forth above, are inconsistent with the tropism theory in its original form, on the basis of the modification of this theory, set forth in the last paragraph? While in the infusoria the assumption of nervous connections, etc., is inadmissible, we may waive that objection and answer the question proposed from an analysis of the observed phenomena. In Stentor or in Stylonychia, for example, we find that the usual reaction to all classes of stimuli is by backing, then turning toward the aboral side; in some of the rotifers by turning toward the aboral (dorsal) side. To simplify matters, we may take into consideration only the turning toward the aboral side. This turning is due to a certain method of movement of certain motor organs. In the rotifers it is the coronal cilia which accomplish the turning, while in the infusoria we know that the adoral cilia are concerned in the movement. We may take the coronal or adoral cilia, then, as representative of the organs active in the turning. For convenience we may designate these active organs simply as x.

Now, when the animal is stimulated on the right side, we find that the motor organs x move in a definite way. On the tropism theory we would conclude, therefore, that the portion of the right side stimulated has nervous connection with the organs x. But we find also that when stimulated on the left side, the oral side, or the aboral side, the organs x move in exactly the same manner. In other words, we find that it does not depend on the side stimulated what organs respond, as required by the tropism theory. This theory, then, in its modified form, is of no more service for these cases than in its original form. The responses in the animals which we have considered must, therefore, be conceived as reactions of the organism as a whole, and due to some physiological change produced by the stimulus, not as the result of direct changes in certain motor organs when they or the parts with which they are most closely connected are locally affected by a stimulating agent. The facts show that the parts act in the service of the whole, not that the action of the whole is due to the more or less independent irritability and activity of the parts.

Thus the facts brought out show that the theory of tropisms is not of great service in helping us to understand the behavior of these lower organisms. On the contrary, the reactions of these organisms seem as a rule thoroughly inconsistent in principle with the fundamental assumptions of the theory.

The facts brought out above are based on a study of what is, of course, a comparatively small number of organisms. They rest chiefly on an extensive study of the ciliate infusoria, with less thorough examination of bacteria, flagellata, rotifers, and a few higher organisms. Doubtless in organisms which are made up of many parts which are less firmly bound together into a unified body than in those considered, we may find greater independence of action in the parts. This seems to be the case, for example, in the sea urchin, with its numerous independently acting spines, pedicellariæ, tube feet, etc. In this animal Von Uexküll (1900, 1900, a) concludes from his extensive study of the reactions that many features in the behavior which seem at first view to be activities of the animal as an individual are really due to the independent reactions of the parts, so that he can say that while in walking, in the case of the dog, " the animal moves its legs ; in the sea urchin the legs move the animal." This method of behavior has a general agreement with what is demanded by the tropism schema, though when we come to details of the behavior of the organs themselves, this theory seems unsatisfactory, even in the sea urchin.

Such organisms as the sea urchin, composed anatomically and physiologically of many parts, each acting almost as an independent animal, are certainly less common than more unified animals, such as we find in the Infusoria, the Rotifera, the flatworms, etc. For this reason, therefore, it has seemed worth while to sum up the real relations of the behavior of these organisms to the tropism theory. The unicellular animals are precisely those on which the prevailing theories of tropisms or taxis have by many writers * been chiefly based. With the demonstration that the behavior of these organisms (as well as of many higher ones), is for the greater part inconsistent with the tropism theory, perhaps a large portion of the foundation for its acceptance as a general formula for the chief features in the behavior of lower animals is cut from beneath it.

In the following paper, on the part played in behavior by physiological conditions of the organism, we shall find other, and, as it seems to me, still more cogent, reasons for holding the tropism theory inadequate to account for the determining factors in the behavior of most lower organisms.

SUMMARY.

The foregoing paper consists of a review of the behavior of Ciliata, Flagellata, Bacteria; of Rotatoria and certain other Metazoa, so far as known from exact observation of their actions when stimulated, with a view to determining how far the prevailing theory of tropisms aids us in understanding the behavior of lower organisms.

The following are considered the essential points in the prevailing theory of tropisms: (1) That orientation is the primary factor in determining the movements of organisms into or out of certain regions, or their collection in or avoidance of certain regions; (2) that the action of the stimulus is directly upon the motor organs of that part of the organism upon which the stimulus impinges, thus giving rise to changes in the state of contraction, which produce orientation.

The reactions of the organisms above named are then reviewed to determine in how far there is agreement with these essential points in the theory of tropisms. The following are pointed out:

The reactions to mechanical stimuli, to chemicals, to heat and cold, and to variations in osmotic pressure have been described in detail, and it is found that orientation is not a primary nor a striking factor in them. The response in all these cases is produced through a " motor

---
* This, however, is not true of Loeb.

reaction" consisting usually of a movement backward, followed by a turning toward a structurally defined side. The direction of turning is thus determined by internal factors.

In the reaction to light orientation is a striking factor, but the orientation is not primary, being due to the production of the same " motor reaction" described in the last paragraph. The method of orientation is incompatible with the idea that orientation is due to the direct action of the stimulus upon the motor organs of the part of the body on which the light impinges, for orientation occurs by turning always toward a certain structurally defined side, without regard to the part of the body struck by the light. The turning may, therefore, be toward or away from the source of light, or in any intermediate direction. In any case it is continued or repeated until the anterior end is directed away from the source of light, when it ceases.

The exact method of reaction to gravity has not been worked out by direct observation.

In the reaction to the electric current the reaction method of the rotifer is by a " motor reflex," and is hence inconsistent with the tropism schema. In the Infusoria there is a partial (but only partial) agreement with the requirements of the tropism theory. But this partial agreement with the theory is due to certain peculiar effects of the electric current which are not known to be produced by any other stimulus. Hence the reaction to the electric current, far from being a type for reactions in general, is a unique phenomenon, demanding special explanation.

The general conclusion is drawn that the theory of tropisms does not go far in helping us to understand the behavior of the lower organisms; on the contrary their reactions, when accurately studied, are, as a rule, inconsistent with its fundamental assumptions. The responses to stimuli are usually reactions of the organisms as wholes, brought about by some physiological change produced by the stimulus; they can not, on account of the way in which they take place, be interpreted as due to the direct effect of stimuli on the motor organs acting more or less independently. The organism reacts as a unit, not as the sum of a number of independently reacting organs.

FIFTH PAPER.

PHYSIOLOGICAL STATES AS DETERMINING FACTORS IN THE BEHAVIOR OF LOWER ORGANISMS.

CONTENTS.

	PAGE
Nature and Evidences of Physiological States,	111
Physiological States in the Protozoa (Stentor as a Type),	112
Physiological States in the Lower Metazoa (the Flatworm as a Type),	115
Changes in the Sense of "Tropisms" and other Reactions,	117
Changes in the Sense of Reactions with Changes in the Intensity of the Stimulus,	118
Interference of Stimuli,	119
Spontaneous Movements,	120
Methods of Causing Changes in Physiological States,	120
Nature of Reactions to Stimuli,	121
Physiological States in the Behavior of Higher Animals as compared with those in Lower Organisms,	124
Summary,	126

PHYSIOLOGICAL STATES AS DETERMINING FACTORS IN THE BEHAVIOR OF LOWER ORGANISMS.

NATURE AND EVIDENCES OF PHYSIOLOGICAL STATES.

In studying the behavior of the lower organisms the units of observation, the factors to which especial attention has been paid have been usually the *tropisms* and *reflexes*. These factors may be considered as determined mainly (1) by the action of external agents on the organism; (2) by the structure of the organism.

An examination of the results of the study of reactions in the lower animals up to the present time shows, I believe, that we must recognize another set of factors in their behavior, of equal importance with either of those already named. This set of factors may be characterized by the general term *physiological states*.

By physiological states we mean the varying internal physiological conditions of the organism, as distinguished from permanent anatomical conditions. Such different internal physiological conditions are not directly perceptible to the observer, but can be inferred from their results in the behavior of the animal. These results are of so marked a character that the inference to different physiological conditions is beyond question.

In the study of tropisms and reflexes a considerable number of instances have been brought to light of changes in the reaction methods, such as can be attributed only to changed physiological conditions. Some of these cases will later be considered in detail in this paper. Comparatively few investigations on the behavior of lower organisms have been published in which attention has been consciously directed to these physiological states, and in most of these the matter has been taken up rather incidentally. We may mention as instances of papers dealing more or less with this aspect of the matter that of Hodge and Aikins (1895) on Vorticella, those of Von Uexküll (1899, 1900, 1900, a, 1903) on the sea urchin and on Sipunculus, my own on the behavior of fixed Infusoria (Jennings, 1902), and that of Pearl (1903) on the flatworm. In the study of higher organisms attention has of necessity been largely directed to the phenomena determined by varying physiological states, as these play a striking part in the behavior.

In the present paper an attempt will be made to collect and analyze a number of the known cases showing the influence of physiological states on the behavior of the lower animals, pointing out some of their bearings on the theories of behavior.

PHYSIOLOGICAL STATES IN THE PROTOZOA (STENTOR AS A TYPE).

We will take up first the lowest organisms in which the matter has been studied in detail, that is, the unicellular animals. These are of special interest in view of their entire lack of a nervous system. As the best-known case we may take the behavior of Stentor. This has been described in full in a former paper by the present author (Jennings, 1902, *a*); for details this paper may be consulted.

When a quiet, extended Stentor is stimulated by lightly touching it or the support to which it is attached with a rod, it reacts by giving a definite reflex, that is, by contracting into its tube.

After this has taken place once or twice we find that the Stentor no longer reacts as before. All the external conditions remain the same; the stimulus applied is the same. Nevertheless, the Stentor does not react. Therefore, we must conclude that the Stentor itself has changed. Its physiological condition is now different from what it was originally. What the nature of the change in its condition is we do not know, save in the fact that the Stentor in this second condition does not react as does the Stentor in the first condition. For the sake of convenience we may number the different physiological conditions, in order that we may determine, if possible, how varied these conditions are. We will call the physiological condition of the undisturbed extended Stentor, before the stimulation, No. 1, or the first condition. The condition in which the Stentor no longer responds to the slight stimulus we will call No. 2.

We may frequently distinguish still a third condition in the behavior under this simple stimulus. At first the Stentor reacts by contraction (condition 1). Then it no longer reacts (condition 2). Later, or in other cases, it may react to the stimulus, but by a different method from the first reaction. It now bends over to one side when touched with the rod. As set forth in my previous paper, "The impression made on the observer is very much as if the organism were at first trying to escape a danger, and later merely trying to avoid an annoyance." As conditioning this third method of behavior, when all outward conditions are the same, we must postulate a third physiological state differing from the other two; this we may call condition No. 3.

We may thus distinguish at least three different physiological states in the reactions to very weak stimuli, where the initial marked response becomes a weak one or disappears. We may now analyze in the same way the behavior under stimuli of a different character, when there is a series of reactions which may be considered of increasing rather than of decreasing intensity. Such a case is that described in my previous paper, when water mixed with carmine particles is allowed to reach the disk of Stentor. The first physiological condition is again No. 1—that of the undisturbed extended Stentor. In this condition the organism does not respond to the stimulus at all. After the stimulus has continued for some time, the organism does respond by turning into a new position. We have, therefore, a new physiological condition. The reaction in this case is the same as that given in condition No. 3, described above. Whether the condition now existing is the same as in the former case we do not know; as we have no positive evidence to the contrary, we will number it 3 also.

Next, after several repetitions of this reaction, the organism responds in a still different manner, by momentarily reversing the ciliary current. Since the stimulus and other external conditions remain the same, the organism itself must have changed. We may call its physiological condition at the present time No. 4.

Next, the animal contracts strongly and repeatedly. This is clearly the result of a still different physiological condition which we may call No. 5.

After thus contracting repeatedly we find that the organism remains contracted much longer than it did at first. It is thus now in a new physiological condition, which we may designate as No. 6.

Finally, it breaks its attachment to the bottom of the tube and swims away through the water. Probably, therefore, we should distinguish a seventh physiological state, corresponding to this reaction. It is possible, however, that the breaking of the attachment is due to the strong contractions which characterize condition No. 6, so that the evidence for a seventh physiological condition is not unmistakable, and it may be omitted from consideration.

We are able to distinguish clearly, therefore, in the study of these two sets of reactions, at least six different physiological states. In each of these states Stentor is a different organism, so far as its reactions to stimuli are concerned. Clearly, then, the external stimuli and the permanent anatomical configuration of the body are by no means the deciding factors in the behavior. These factors, in the reaction series last described, permit at least five different methods of behavior. Which of these methods is actually realized depends not on the quality or intensity of the stimulus, nor on the anatomical structure of the organism, but on its physiological condition.

I do not wish to imply that I hold that the six different physiological states above distinguished are sharply defined, separate things. On the contrary, it is much more probable that the different physiological conditions form a continuum. We can, by taking sections, as it were, at different intervals, distinguish at least six actually different conditions; but doubtless there exists every possible gradation from one to another, so that still other actually different conditions could be distinguished if we had criteria for separating them. By careful analytical experiments the number of different physiological conditions clearly recognizable could doubtless be increased.

In other unicellular organisms doubtless a condition of affairs may be found similar to that set forth above for Stentor. Hodge & Aikins (1895) showed that the reactions of Vorticella vary with its physiological condition. In the same paper (Jennings, 1902, *a*) in which the behavior of Stentor was described, I have shown that in various other fixed infusoria (Carchesium, Epistylis, etc.) the behavior likewise depends upon physiological states of the organism. In the free-swimming infusoria this has not been shown to be true, at least to any such extent. There may two grounds for this. Firstly, it is probable that in the free-swimming infusoria the behavior is actually less varied than in the fixed species. A single motor reaction usually removes them from the action of the stimulus causing it, so that there is no reason for a recourse to other methods of reaction, as occurs in Stentor. Secondly, in the free-swimming infusoria it is difficult, almost impossible, to observe continuously the reactions of a given single individual. This difficulty could doubtless be met by proper methods of experimentation, and if this were done it can hardly be doubted that a dependence of the reactions on the physiological states of the organism would be found here also. Indeed, we have indirect evidence that this is true in the case of Paramecium, in work already published. Thus, in one of my earlier papers (1899, *a*, p. 374) I called attention to the fact that Paramecia from different cultures often vary exceedingly in their reaction to a given solution of a chemical. Still more pertinent to the point under consideration is the fact, described in the first of my studies (Jennings, 1897), as well as in the recent paper of Pütter (1900), of the great difference in the reaction of Paramecia and other fixed infusoria when in contact with a solid, as contrasted with their reaction when not thus in contact. As this, however, may be interpreted as an interference of two stimuli, a discussion of the point is reserved until later. A study of individual specimens of some of the larger Hypotricha, such as Stylonychia, from the point of view of changes in reaction methods with changes in physiological condition, would doubtless bring forth interesting results.

Even in the lower unicellular organisms, the Rhizopoda, similar dependence of the method of reaction on the physiological state of the organism is known to exist. Thus Rhumbler (1898, p. 241) has observed that *Amœba verrucosa* may at first begin to ingest an Alga filament, then later, before the ingestion is complete, the filament may be ejected. This involves a change in the physiological condition of the Amœba; otherwise it would not now reject a certain object which it before ingested.

PHYSIOLOGICAL STATES IN THE LOWER METAZOA (THE FLATWORM AS A TYPE).

Passing now to the Metazoa, we find in the flatworm, Planaria, as described by Pearl (1903), a dependence of the reaction of the organism on its physiological condition similar to that which we saw above for Stentor. The flatworm may be considered typical of the lower bilateral Metazoa, so that it will be worth while to subject some features in its behavior to a brief analysis from our present point of view.

We may examine for a simple typical case the reactions to mechanical stimuli applied to one side of the anterior part of the body. The flatworm is touched on one side with the tip of a hair or of a fine glass rod. The resulting response is one of two reactions—the worm turns either toward the point stimulated (positive reaction) or away from it (negative reaction). Typically, the positive reaction is given to a weak stimulus, while the negative reaction results from a strong stimulus. The words "weak" and "strong" have, as we shall see, only a relative meaning when used in this connection.

When, now, we ask which of these reactions shall be given as a response to a certain stimulus, we find that this depends upon the physiological condition of the organism. Pearl finds the reactions determined by the following definitely marked physiological states:

1. Individuals are frequently in what may be called a *resting* condition. The tonus is lowered; the animals are very inactive and do not respond readily to stimuli. This condition is compared by Pearl with that of sleep in higher animals. When a flatworm in this condition is given a light stimulus, such as would in an active specimen induce a positive reaction, it does not respond at all. If the strength of the stimulus is increased, the animal finally responds with a negative reaction, turning away from the point stimulated. We may call this the condition of lowered tonus.

2. In the more usual active condition the flatworm gives the positive reaction to a very light touch, a negative reaction to a stronger stimulus. We may call this the normal condition.

3. After the animal has been repeatedly stimulated it seems to become excited; it moves about rapidly, and now gives always the negative reaction to any mechanical stimulus to which it reacts at all. It behaves much as many higher animals do under the influence of fear. We may call this the excited condition.

4. After the worm in the excited condition has been stimulated repeatedly on one side, so that it turns its head steadily in the opposite direction, after a time it suddenly changes its method of reaction. It jerks backward, then turns the anterior end quickly through a considerable arc, usually toward the side from which the stimulus is coming, so that the head now points in an entirely new direction. Since the stimulus and other external conditions remain the same, the organism must have passed into a new physiological condition, or it would not now react in a different way. We may call this for convenience the condition of over-stimulation.

5. Sometimes individuals are found which for a brief period (two or three hours) seem in a much more active condition than usual. They move about rapidly, but do not conduct themselves like the excited individuals. As they move they keep the anterior end raised and wave it continually from side to side as if searching. Specimens in this condition react to almost all mechanical stimuli, whether weak or strong, by the positive reaction, turning toward the point stimulated. Experimentation failed to show that this condition was due to hunger. We may speak of this fifth physiological condition as the state of heightened activity.

In addition to the effect of these five well-defined physiological states on the method of reaction to mechanical stimuli at the anterior part of the body, Pearl finds that other less easily definable internal conditions affect the reactions. At times an individual will give the positive reaction to a stimulus of a certain strength a few times, then cease to give it. On account of these and other complications due to varying internal conditions, Pearl concludes:

It is almost an absolute necessity that one should become familiar, or perhaps better, intimate, with an organism, so that he *knows* it in something the same way that he knows a person, before he can get even an approximation of the truth regarding its behavior.

We have taken up above only the physiological conditions influencing the reactions to simple mechanical stimuli in the anterior region of the body. We find the condition of affairs even here somewhat involved. When other more complex stimuli are taken into consideration the results of the interplay of change of physiological condition and variations in the stimuli become, of course, much more complicated.

Thus we find in the bilateral metazoan Planaria, as in the unsymmetrical protozoan Stentor, that we can by no means predict the behavior of the individual from a knowledge of the anatomical structure and of the strength of the stimulus. The anatomical structure limits the possibilities of reaction to several methods, which are, however, entirely different or opposite in their effects on the relation of the organism to the stimulus. Just which of these reactions shall be given as a response to any particular stimulus depends on the physiological condition of the organism. This physiological condition depends largely, as we shall note later, on the history of the individual. Thus no single fixed schema, such as we have in the tropism theory, can ever possibly explain or define the essential points in the behavior of an animal.

Stentor and Planaria may be taken as typical examples of the higher Protozoa and of the lower Metazoa, respectively. It is true that we are not so well informed as to changes in physiological condition in other lower organisms as in these two cases, but this is unquestionably due merely to the fact that investigation has not been directed especially to this point. There are, however, many cases in the literature which explicitly or implicitly show the importance of physiological conditions in determining the behavior of lower organisms. A number of these cases may be brought together here.

CHANGES IN THE SENSE OF "TROPISMS" AND OTHER REACTIONS.

Loeb (1893) and Nagel (1894) have called attention to the fact that certain worms and mollusks respond to a shadow by suddenly withdrawing into their tubes, but that after the first reaction has been thus produced the worms may no longer react. In this "after effect of the stimulus" (Loeb) we have, of course, a case of changed physiological condition.

Changes in the sense of "tropisms" belong here. Groom & Loeb (1890) found that larvæ of Balanus are at certain times of the day positively phototactic; at other times negatively phototactic. This difference, in so far as it is independent of changed external conditions, is, of course, due to differences in the physiological condition of the organism. Loeb (1893) found that the larvæ of the moth Porthesia are positively phototactic when hungry; not so after eating. Here we have a well-defined physiological condition determining the nature of the reaction; hunger is one of the most important conditions in many of the lower animals. Sosnowski (1899) and Moore (1903) show that the geotropism of Paramecium changes from negative to positive under various conditions. Towle (1900) and Yerkes (1900) have shown that the sense of the phototactic reaction in Entomostraca is dependent on the preceding treatment of the organism, mere transference with the pipette often changing the sense of the reaction from positive to negative or *vice versa*. Such instances could doubtless be multiplied indefinitely. Each case taken by itself seems perhaps of comparatively little significance. We may look upon them, however, as indications of an extensive dependence of behavior on physiological conditions, such as we found in Stentor and the flatworm. Thorough investigation of any of these organisms from this point of view would doubtless bring to light a variety of physiological conditions on which the reactions depend.

CHANGES IN THE SENSE OF REACTIONS WITH CHANGES IN THE INTENSITY OF THE STIMULUS.

Must we not bring under the same point of view the well-known phenomenon of a change in the sense of the reaction with a change in the intensity of the stimulus? As a simplest case of this we may take the reaction of Stentor to mechanical stimuli. As shown in the ninth of my studies (Jennings, 1902), Stentor reacts to a very weak mechanical stimulus on one side of the disk by bending toward the source of stimulus; to a stronger but otherwise similar stimulus it responds by contracting into the tube, or (later) by bending in another direction. In the same way the flatworm reacts positively to a weak mechanical or chemical stimulus, negatively to a stronger one. How can we explain these opposite reactions to stimuli of the same quality, differing only in intensity?

We have here, it seems to me, the same phenomenon shown in the production of a change in physiological condition by a stimulus. We know that even a single stimulus may produce a changed physiological condition, as when after a single stimulus the organism no longer reacts as before. We know also that the nature of the physiological condition determines the reaction. In the present case we must conclude that a light stimulus throws the organism into a certain physiological condition, whose concomitant reaction is turning toward the point stimulated. A more intense stimulus induces a different physiological condition, whose concomitant reaction is a contraction into the tube (Stentor), or a turning in the opposite direction (flatworm). The action of the stimulus, as we have seen in the foregoing paper devoted to the theory of tropisms, cannot in most cases be directly on the motor organs, so that from this point of view also we are almost forced to the conclusion that the primary action of the stimulus is to change the physiological condition of the organism. In any reaction to stimulus we would have, therefore, the following steps: The stimulus acting on the organism changes its physiological condition; this physiological condition induces a certain type of reaction. In determining what physiological condition shall be produced, the intensity of the stimulus is fully as important as its quality.

We have a similar reversal of the reaction as the intensity changes in reactions to light. Many organisms are positive to weak light; negative to strong light. The cause of this reversal of the reaction as the light grows stronger has given rise to much discussion (see Holmes, 1901 and 1903). We have here, of course, a parallel case to the reversal of the reaction in Stentor or the flatworm under mechanical stimuli of varying intensity. In the weak light we must suppose the organism to be thrown into a certain physiological condition, the concomitant of which is a certain type of reaction. In a more intense light a different physiological condition is induced, corresponding to a different reaction. The fact that different intensities of stimuli do cause different physiological conditions and different reactions is, of course, familiar to us, both from experimentation on animals and from our own experience; in the latter case we usually call the distinctive reactions to very intense stimuli pain reactions. In the reversal of the reaction to light as the light becomes stronger we have, it seems to me, merely an instance of this general phenomenon, not differing in fundamental character from other instances.

In most of these cases we have, of course, a further problem in regard to those features of the reaction which concern direction. Why does the weak stimulus on the left side of Planaria cause a turning toward that particular side? Or, why does a weak light from a certain direction cause Volvox to swim in that particular direction? These problems of direction are, of course, not touched in the foregoing discussion, which, however, loses none of its force because these problems remain. They are simply farther problems. The tropism theory gave a simple, direct answer to these questions; but, as we have already shown in the foregoing paper, this answer was, in many cases at least, not a correct one. Possibly some combination of certain features of the tropism theory with a consideration of the facts of changes in physiological condition may give us a satisfactory answer to the problems of direction.

INTERFERENCE OF STIMULI.

Again, we have in the interaction or interference of stimuli certain phenomena which seem to fall under our present point of view. First, we have the so-called cases of "heterogeneous induction," where the action of one stimulus reverses or modifies the reaction to another. For example, many cases are known in which animals positively phototactic become negative, or *vice versa*, when the temperature is changed. (For a collection of such cases see Loeb, 1893, and Davenport, 1897, p. 199.) In these cases the physiological condition of the organism seems altered by one stimulus (as heat or cold), in such a way that it no longer reacts to another stimulus (light) as it did before. In the ciliate infusoria, specimens which are in contact with solids do not react at all to many agents which under other circumstances call forth a marked reaction. Pütter (1900) has made a special study of this matter, describing especially the interference of contact with the reaction to heat and to electricity. In the second of these contributions (p. 32), we have seen that attached Stentors do not react at all to light. Physically considered, there is no necessary opposition between the action of contact and the action of the other stimuli named. We must conclude then that contact with solids so alters the physiological condition of the organism that it no longer reacts to the other stimuli.

SPONTANEOUS MOVEMENTS.

Further, we find that alterations in physiological condition may cause definite movements, which take place without external stimulus. Such are the movements which we call spontaneous. As an example of this we may take the case of Hydra. If an undisturbed green Hydra is observed continuously, it is found to contract and again to extend without visible cause every $1\frac{1}{2}$ to 2 minutes. Thus, it remains at rest for a period of, say, $1\frac{1}{2}$ minutes. Its physiological condition at this time we may call X. At the end of this period it contracts. Since the external conditions have not changed the Hydra itself must have changed, otherwise it would continue at rest. The physiological condition X passes into the condition Y, and the Hydra as a result contracts. This contraction is, of course, exactly the reaction given as a response to most stimuli in Hydra. In Vorticella we find similar spontaneous contractions at intervals, essentially as in Hydra. Cases of movements in the lower organisms that are inaugurated by internal changes in condition could, of course, be multiplied indefinitely. For our present point of view it is of importance to recognize clearly the fact that a change in physiological condition may, by itself, cause exactly the same behavior that at other times appears as a response to external stimuli.

METHODS OF CAUSING CHANGES IN PHYSIOLOGICAL CONDITION.

Changes in physiological condition are thus evidently brought about in a number of different ways. We may attempt to summarize here the different methods which appear to exist in the lower organisms.

(1) A single simple stimulus may bring about a change in physiological condition. This is proved by the fact that the organism after it has received a single stimulus may react differently from its previous method. Thus, Stentor reacts to a single touch, but *after* this single touch it may no longer react when touched in the same way again; or it may react in a different manner. It is probable, further, that the first reaction to a single simple stimulus is to be considered due to a change in physiological condition produced by this stimulus.

(2) Repetition of the same stimulus may cause a change in physiological condition such as is not produced by a single stimulus. This

again may be illustrated from the experiments on Stentor, described above (p. 112).

(3) Internal causes, not definable, may give rise to changes in physiological condition. The result is spontaneous movement at intervals, as described above for Vorticella and Hydra. As many authors have pointed out, rhythmical spontaneous movements may be due to a steady, non-rhythmical, internal change of condition.

(4) The movement or reaction performed by the organism may change the physiological condition. This is illustrated in one way by the fact that after a single spontaneous contraction in Vorticella or Hydra, the animal remains quiet for an interval, showing that the original physiological condition was restored by the movement. The fact that the reaction performed by the organism changes the physiological condition of the latter is, of course, the basis of the formation of *habits* in higher organisms; in this case the performance of a reaction once or repeatedly throws the organism into a condition where it is more likely to react in the same way again. This particular method of alteration of condition has perhaps not been clearly demonstrated for unicellular organisms, though there is some indication of it in the behavior of Vorticella, as described by Hodge & Aikins (1895), and of Stentor as described by myself in the ninth of my studies. Thus, Stentor responds to carmine in the water by a series of different reactions, finally reaching the condition where it reacts by contracting at once into its tube. If the stimulus is now repeated every time the Stentor extends, it never gives its earlier method of reaction, but reacts steadily for a long time by contracting at each stimulus. Is this persistence in the contraction reaction due partly to the fact that it has begun on this reaction method, and therefore keeps it up, or is it due only to the fact that the stimulus has been repeated many times? In the former case the behavior would perhaps fall under our present point of view; in the latter it would not. Cases among the Protozoa where the repeated performance of a reaction clearly makes the further performance of the same reaction easier or more likely to occur, would be of much interest.

NATURE OF REACTIONS TO STIMULI.

The foregoing considerations evidently have a definite bearing on the problem of the nature of reactions to stimuli. They lead, as set forth briefly on page 118, to the following conception of the steps occurring in a reaction to a stimulus: (1) The stimulus acting on the organism causes a change in its physiological condition; (2) this change in physiological condition gives rise to the typical reaction.

The evidence for this view is found scattered throughout the foregoing discussion; its main points may be briefly summarized here as follows:

(1) We have seen that stimuli do unquestionably cause changes in physiological condition. This is demonstrated by the fact that after a stimulus has occurred and *ceased* to act, the organism reacts differently to the same or other stimuli. (For examples see pages 112, 117.)

(2) We have seen that the changes in physiological condition do unquestionably cause definite movements, of exactly the sort that we are accustomed to call reactions to stimuli. (Contraction of Hydra or Vorticella, etc.; see p. 120.)

These two facts give a solid foundation for the above view of reactions to stimuli, and, indeed, it seems to me, raise a presumption that reactions to stimuli are, as a rule, brought about in the way described. Further evidence in favor of this view is as follows:

(3) In the paper which precedes the present one we have demonstrated that, in the Infusoria and Rotifera at least, the action of stimuli is not directly on the motor organs of that part of the body on which the stimulus impinges. The organism reacts as a whole, and in a way that is not explicable even on the assumption of a definite plan of nervous interconnection between the regions stimulated and the motor organs, an assumption that is, of course, in any case not allowable for the Infusoria. Such reactions cannot be explained otherwise than as due to changes in the physiological condition of the organism as a whole. Further, evidence was given to show that the reactions of higher organisms are in many cases equally inexplicable as a result of direct action of the stimulus on the motor organs.

Only in the reaction of some organisms to the constant electric current did we find such conditions fulfilled as permit an explanation of a part of the phenomena on the theory of the direct action of the agent on that part of the body on which it impinges, in accordance with the theory of tropisms. Other features of the reaction to this stimulus (in many cases the determining ones) are only explicable on the theory that they are due to the physiological state of the organisms as a whole, induced by the stimulus (see p. 100). This shows that we may find at any time these two methods of action mixed, or perhaps either one separately. But the reaction to the electric current is the only one out of the reactions to a multitude of agents, that, in the Infusoria, has been shown to have this additional feature—reactions of different parts of the body in opposed ways. The reaction to the electric current is thus of the very greatest interest, not because it stands as type for reactions in general, but for exactly the opposite reason, because it presents factors which are not known to occur in other reactions.

(4) The view that reactions to stimuli take place through the intermediation of changes in the physiological condition of the organism as a whole is further reinforced by the fact, set forth above, that it is only

on the basis of such a view that we can understand the changes of reaction which occur when the same stimulus is repeated ; by the facts of the interference of stimuli, even when their direct physical action is by no means opposed ; by the facts of heterogeneous induction, and by the fact that organisms at different times of the day or at different seasons show different methods of reaction to the same stimuli.

(5) This view is also strengthened by the fact that it brings into relation reactions to stimuli and spontaneous movements. On this view both are due directly to the same cause, to changes in physiological condition, produced in one case by internal causes, in the other by external causes.

(6) This view receives powerful support, it seems to me, in our knowledge of what takes place in the higher animals, including man. This point I shall attempt to develop farther on in the present paper.

This view, that reactions to stimuli in the lower organisms are produced in general through changes in physiological condition, is not, of course, set forward as anything new or original. Many others have doubtless taken this point of view, and it is implied, perhaps not always consciously, in many attempted explanations of animal behavior. The writer is merely attempting to emphasize that particular interpretation, out of many existing ones, towards which the facts seem to point strongly. He is convinced that the factor of physiological condition as determining behavior has not been so fully and explicitly realized and dealt with in work on the lower organisms as the facts demand, and that many things that seem anomalous fall into their proper places when this factor is taken fully into consideration.

What is the nature of physiological conditions, or changes in physiological condition? Of course, we are not able to answer this question. One is tempted to think of these expressions as signifying something like chemical states or changes in chemical states. But the concept of physiological states is, for higher animals at least, one at which we arrive by analysis of complex phenomena in behavior, and this does not give us any direct evidence as to the real nature of the change in the living substance (considered as matter) which takes place when the physiological condition changes.

The concept " physiological states" is a preliminary collective concept, which may later be analyzed into many. Such analysis is certain, however, to be difficult and hypothetical in character in the lowest organisms. In man we have, of course, a basis for analysis in the subjective accompaniments of physiological (here called psychological) conditions,—in the feelings, emotions, etc.

PHYSIOLOGICAL STATES IN BEHAVIOR OF HIGHER ANIMALS, AS COMPARED WITH THOSE IN LOWER ORGANISMS.

Realization of the fact that the behavior, even in the lowest organisms, is determined to a large degree by physiological states must be of great service in welding into one connected whole the study of behavior in all animals, from the lowest up to man. The attempt to divorce the study of the behavior of man from that of the lower animals, which has been evident in late years, seems unfortunate and unnecessary. It is true that we are not justified in reading the subjective states of man directly into the lower organisms. But we are not confronted with the alternative of doing this or of separating the two subjects completely. The behavior of man can be studied from the same objective standpoint which we employ in investigating the behavior of animals. When this is done, there is no reason for holding the results on man aloof from those obtained elsewhere; if it is proper to compare different organisms of any kind from this point of view, in order to obtain general results, as all investigators do, it is certainly proper to draw man also into the circle of comparison. The fact that in man we can know also the subjective accompaniments of the different physiological states and reactions is by no means a disadvantage in this comparison; it is merely an additional feature, of the highest possible interest. We can even, it seems to me, justifiably call attention to the relation between the subjective states as found in man to certain general phenomena common to man and other organisms. It is only when we proceed directly to attribute to the lower animals the subjective states which we know only in man (and, indeed, only in our own individual minds) that we pass the boundary of scientific procedure.

In the higher animals, and especially in man, the essential features in behavior depend very largely on the history of the individual; in other words, upon the present physiological condition of the individual, as determined by the stimuli it has received and the reactions it has performed. But in this respect the higher animals do not differ in principle, but only in degree, from the lower organisms, as we have seen in our analysis of the behavior of Stentor. In this unicellular form we were forced to distinguish at least six different physiological conditions, determining in the same individual different reactions to the same stimuli. In the higher animals, and especially in man, we can distinguish, as might be expected, an immensely greater number of such conditions which induce different reactions, but there is no evident difference in principle in the two cases. Can we go farther and make a more direct comparison of individual physiological states in the higher and lower organisms? We find in Stentor, and again in the flatworm, that after the organism has been repeatedly stimulated by an agent which must in the long run be classed as injurious, it is thrown into a physiological condition in which its reactions become more rapid and powerful, and of such a nature as to remove the organism from the source of stimulus. We find that in this state the organism reacts to any stimulus to which it reacts at all by a strong negative reaction. In higher animals we frequently find the same condition of affairs, and the animal is then commonly said to be frightened. Finally, we often find in man a similar condition, and here we know certain subjective accompaniments of the physiological condition, the most characteristic of which is perhaps the emotion of fear. In all these cases the objective manifestations of the physiological condition are of the same character. Does the fact that in man we know something additional about the matter, the subjective accompaniments, constitute grounds for denying the essential similarity, from a physiological standpoint, of this condition in man and that in the lower organisms? It seems to me that it does not; in fact, all that is maintained in making the comparison is that this condition causes similar objective phenomena and is brought about by similar conditions. Further than this our analysis and comparison cannot go.

Another class of physiological conditions which we can distinguish almost all the way through the animal series is that produced characteristically by intense stimuli, as opposed to faint stimuli. As a rule, any stimulus, even if it is one to which the organisms respond usually by a positive reaction, produces, when it becomes very intense, reactions whose general effect is to remove the organisms from the source of stimulation (negative reactions).* This is true in Amœba, where weak mechanical stimuli cause spreading out and movement toward the source of stimulus, while strong mechanical stimuli cause it to contract and move away; it is true for Stentor; it is true for the stimulus of weak and strong light in Euglena and Volvox and many other organisms; it is true for mechanical and chemical stimuli in the flatworm; it is true in general for higher animals and man. In all these cases the intense stimulus evidently changes the physiological condition so that the organism now reacts negatively. In man we know that this physiological condition is accompanied subjectively by pain or at least discomfort, and even in higher animals such reactions are usually spoken of as pain reactions. Objectively considered, the phenomena are analogous throughout the animal series, so that we

---

* "All organisms behave in two great and opposite ways toward stimulations; they approach them or they recede from them. Creatures which move as a whole move toward some kinds of stimulations, and recede from others. Creatures which are fixed in their habitat expand toward certain stimulations, and contract away from others." Baldwin, 1897, p. 199.

may properly characterize the physiological condition which produces the negative reaction to strong stimuli, with Professor J. Mark Baldwin (1897, p. 43) as the "physiological analogue of pain." This, of course, by no means commits us to the belief that the organisms have a *sensation* of pain; concerning this we know nothing.

It thus seems to me possible to trace some of the physiological conditions which we know, from objective evidence, to exist in man and the higher animals, back to the lowest organisms. Many conditions that we can clearly distinguish in man will doubtless be followed back to a common single condition in the lower organisms; but this is exactly what we should expect. Differentiation takes place as we pass upward in the scale, in these matters as in others.

The most interesting and important field in which we find the behavior of higher organisms dependent on their previous history, and, therefore, on their present condition as influenced by previous experience, is in that group of phenomena which we call memory, or learning by experience. Memory has as its basis the general phenomenon that a stimulus received or a reaction performed leaves a trace on the organism, or modifies its condition in such a way that it later reacts differently to the same stimulus. This basis of memory is, of course, clearly present in Stentor.

The analysis of the different physiological conditions found in the lower organisms, the influences to which they are due, and the reactions of these organisms as influenced by physiological conditions certainly forms a most promising field for research, and one as yet almost untouched.

SUMMARY.

The present paper attempts to show, by an analysis of certain phenomena in the behavior of lower organisms, taking Stentor and Planaria as types, that physiological states of the organism are most important determining factors in reactions and behavior. In these organisms, to the same stimuli, under the same external conditions, the same individuals react at different times in radically different ways, showing the existence of different physiological states of the organism, which determine the nature of the reactions. In a unicellular organism (Stentor) we can distinguish at least six different physiological states, in each of which the organism has a different reaction method, and corresponding facts are brought out for the flatworm. Scattering observations taken from works on tropisms, etc., are shown to indicate that the same state of affairs is found in other lower organisms.

The conditions producing these different physiological states are examined and their importance for the theory of behavior in the lower organisms is brought out. The relations of these facts to "interference of stimuli," "heterogeneous induction," "spontaneous movements," and "changes in the sense of reactions with a change of intensity in the stimulus," are developed.

The view is set forth that in most of the lower organisms a reaction to stimulus usually involves the following factors: (1) the stimulus changes the physiological state of the organism as a whole; (2) this change in physiological state induces a certain type of reaction. Evidence for this view is summarized.

Finally, it is pointed out that realization of the importance of physiological states as determining factors in the behavior of the lower organism is of service in bringing the study of these organisms into relation with that of higher animals and man. An objective study of the behavior of these higher animals shows the prevalence of physiological states as determining factors in behavior, and in some cases, at least, some of these states are closely analogous to what we find even in unicellular organisms.

SIXTH PAPER.

THE MOVEMENTS AND REACTIONS OF AMŒBA.

CONTENTS.

	PAGE.
Introduction: Objects of the Investigation,	131
Description of the Movements and Reactions,	132
The Movements,	132

    The Movements of Amœba as described by Rhumbler and Bütschli; Agreement with Currents in a Drop of Fluid Moving as a Result of a Local Decrease in Surface Tension.......... 132
    Currents in Amœba as studied from above; Lack of Backward Currents........ 134
    Movements of Upper and Lower Surfaces Studied Experimentally; Rolling Movement.................................. 138
    Amœba verrucosa and Its Relatives. 140
    Other Species of Amœba................ 146
    Historical on Rolling Movements in Amœba 148

    Formation and Retraction of Pseudopodia 152
    Surface Currents in Formation of Pseudopodia in contact with Substratum 152
    Formation of Free Pseudopodia........ 153
    Withdrawal of Pseudopodia............. 156
    Movements at Anterior Edge............ 160
    Movements of Posterior Part of Body.... 165
    General View of Movements of Amœba in Locomotion 169
    Some Characteristics of the Substance of Amœba........................... 173
    Fluidity.................................. 173
    Rhumbler's Ento-ectoplasm Process 173
    Elasticity of Form in Amœba........... 175
    Contractility in Ectosarc of Amœba.. 177

   Reactions to Stimuli, ... 181

    Reactions to Mechanical Stimuli............ 181
      Positive Reaction.................... 181
      Negative Reaction.................... 182
    Reaction to Chemical Stimuli.............. 187
    Reaction to Heat......................... 190
    Reactions to Other Simple Stimuli......... 191

    Some Complex Activities.................. 193
    Activities connected with Food-taking 193
      Taking Food........................ 193
      Pursuit of Food.................... 196
      Other Amœbæ as Food.............. 198
    Reactions to Injuries..................... 202

Physical Theories and Physical Imitations of Amœboid Movements, . 204

    Surface Tension Theory..................... 204
    Berthold's Theory that One-sided Adherence to Substratum is the Cause of Locomotion 208
    Experimental Imitation of Locomotion in Amœba.................... 209
    Formation of Free Pseudopodia....... 214

    Experimental Imitation of Movements due to Local Contractions of Ectosarc and of the Roughening of Ectosarc in Contraction 215
    Direct or Indirect Action of External Agents in Modifying Movements....... 219
    Direct or Indirect Action in Food-taking, 222
    General Conclusion....................... 225

Behavior of Amœba from Standpoint of Comparative Study of Animal Behavior, ... 226

    Habits in Amœba....................... 226
    Classes of Stimuli to which Amœba Reacts 227
    Types of Reaction...................... 227

    Relation of Different Reactions to Different Stimuli; Adaptation in Behavior of Amœba 227
    Reflexes and "Automatic Actions" in Amœba 228
    Variability and Modifiability of Reactions 229

Summary, ... 230

THE MOVEMENTS AND REACTIONS OF AMŒBA.

INTRODUCTION: OBJECTS OF THE INVESTIGATION.

The present paper contains the results of an investigation which was undertaken with two general problems in mind. The first purpose was to determine by observation and experiment, from the standpoint of the student of animal behavior, how far recent physical and mechanical theories go in aiding us to explain the behavior of Amœba. The second object of the work was to furnish needed additional data on the reactions of Amœba to stimuli, and to systematize and unify our knowledge of its behavior.

The recent theories which would resolve the activities of Amœba largely into phenomena due to alterations in the surface tension of a complex fluid seem to promise much. They are of precisely the character from which most may be hoped; from a study of the physics of matter in a state similar to that found in the living substance, the laws of action of this living substance are sought. Such theories have been developed, as is well known, by Berthold (1886), Quincke (1888), Bütschli (1892), Verworn (1892), Rhumbler (1898), Bernstein (1900), Jensen (1901), and others. The success of this method of attacking the problems seems great. Activities similar, at least externally, to those of Amœba, are produced by physical means, and fully analyzed from the physical and mechanical standpoint. In this manner the movement, the control of movement by external agents, the feeding, the choice of food, the making of the shell, and other features of the behavior have been more or less closely imitated,[*] and in a way permitting a complete analysis in accordance with chemical and physical laws.

From the standpoint of the student of animal behavior, the resolution of the behavior of any organism into the action of known physical laws must be a matter of the deepest interest. The actions of higher organisms seem at present so far from such a resolution that some investigators believe an essential difference in principle to exist between the behavior of living things and non-living things; between the laws of biology and those of physics. The resolution, then, of the behavior of even the simplest organism into known physical factors would be an event of capital significance, affecting fundamentally the whole theory of animal behavior. A renewed thorough study of the

---

[*] See especially Rhumbler, 1898.

131

facts, with especial reference to these theories, seems, therefore, much to be desired. The results of the present study will show, I believe, that such a re-examination of the facts was greatly needed.

As to the second object of this investigation, stated above, it is a somewhat remarkable fact that the observational basis for a number of the most important reactions assumed to exist in Amœba is exceedingly scanty, particularly so far as control of the direction of movement is concerned. For example, one of the reactions most often assumed to exist in Amœba, and most commonly selected for imitation by physical means, is chemotaxis, the movement toward or away from a diffusing chemical. But no account exists, so far as I have been able to discover, of actual observation of such a reaction in Amœba, under experimental conditions. Again, the effects of slight or of intense localized mechanical stimuli, in controlling the direction of movement, has not been worked out in detail. To fill these and similar gaps in our knowledge, and to bring the different reactions into relation with each other, so as to make possible a connected account of the behavior of Amœba, is, then, the second object of this paper.

I shall first give an account of the movements and reactions of Amœba, as determined by observation and experiment, without entering in detail upon the theories of the subject. This will be followed by a section dealing with the physical theories and physical imitations of the movements and reactions, in the light of the facts set forth in the first section. A brief final section will be devoted to a characterization of the behavior of Amœba from the standpoint of the student of animal behavior.

I am compelled to give a full description of the normal movements of Amœba, as the course of the investigation showed that the prevalent conception of these movements, on which many of the theories have been based, is not correct.

DESCRIPTION OF THE MOVEMENTS AND REACTIONS.

THE MOVEMENTS.

MOVEMENTS OF AMŒBA AS DESCRIBED BY RHUMBLER AND BÜTSCHLI; AGREEMENT WITH CURRENTS IN A DROP OF FLUID MOVING AS A RESULT OF LOCAL DECREASE IN SURFACE TENSION.

There are few subjects that have been studied more than the nature of the movements of Amœba, but nothing final has been reached, even from the descriptive standpoint. The first preliminary to an understanding of the nature of the movements must be to determine just what movements take place.

The most extensive recent study of the movements of Amœba has been made by Rhumbler (1898), though the magnificent monograph of the Rhizopods by Penard (1902) contains incidentally a large number of valuable observations on this matter.

According to Rhumbler (*l. c.*) the movements in normal locomotion are typically as follows: From the hinder end of the Amœba (or of the pseudopodium, if a single pseudopodium is under consideration) a current of endosarc passes forward in the middle axis; in front this flows outward toward the sides, then backward along the surface, gradually coming to rest. Figs. 30 and 31, taken from Rhumbler, give diagrams of these currents in an Amœba moving as a whole (Fig. 30), and in the formation of pseudopodia (Fig. 31). In an Amœba which forms more than one pseudopodium at once, these typical currents become somewhat complicated (Fig. 32), but retain their main features. The backward current shown at the sides in Figs. 30–32 is conceived to be present also above and below, that is, over the whole surface of the Amœba. A diagram of the currents in side view, as given by Rhumbler, is shown in Fig. 33, B. An essentially similar account of the currents is given by Bütschli (1880, 1892).

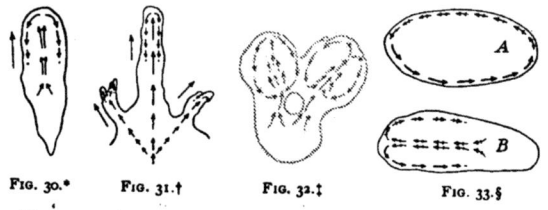

FIG. 30.*   FIG. 31.†   FIG. 32.‡   FIG. 33.§

The most striking feature in the currents as above set forth is the fact that they agree precisely with the currents produced in a drop of fluid of any sort when the surface tension is lowered over a certain limited area. There is always a current over the surface away from the region where the tension is lowered, while an axial current moves toward the

---

* FIG. 30.—Diagram of the currents in a progressing *Amœba limax*, after Rhumbler (1898).

† FIG. 31.—Diagram of the "fountain currents" in pseudopodia of Amœba, after Rhumbler (1898).

‡ FIG. 32.—Diagram of complex "fountain currents" in an Amœba with two large pseudopodia, after Rhumbler (1898).

§ FIG. 33.—Comparative diagrams of the currents in a rolling movement, and in the movement of Amœba, as conceived by Rhumbler, viewed from the side. In *A* are represented what Rhumbler conceives to be the necessary currents in a rolling movement, while *B* represents what Rhumbler considers the really existing currents in Amœba, as seen from the side. The heavier arrows in each case represent the current on the lower surface. After Rhumbler (1898).

point of lowered tension. Diagrams of the movement of such drops are given in Fig. 34. Further, the drop may elongate in the direction of the axial current, and may move bodily in that direction, just as happens in Amœba.* It is most natural, therefore, to conclude as Bütschli (1892) and Rhumbler (1898) have done, that the movements of Amœba are likewise due to a lowering of the surface tension at the anterior end, *provided that its movements really take place in the way described above.*

CURRENTS IN AMŒBA AS STUDIED FROM ABOVE; LACK OF BACKWARD CURRENTS.

At the beginning of my work I had no doubt that the movements occurred exactly as above described, and, therefore, did not devote special attention to this point. But I was soon struck by the fact that I was unable to see any backward current at the sides, as represented in Figs. 30 and 31. Further careful study of the movements of *Amœba limax, A. proteus, A. angulata, A. verrucosa, A. sphæronucleolus,* and one or two undetermined species confirmed this fact, and I may say at once that after several months' continuous study of the movements and reactions of Amœba I have never, except in one or two doubtful instances, seen any backward movement of the substance at the sides or on the surface of an Amœba that was moving forward in a definite direction.

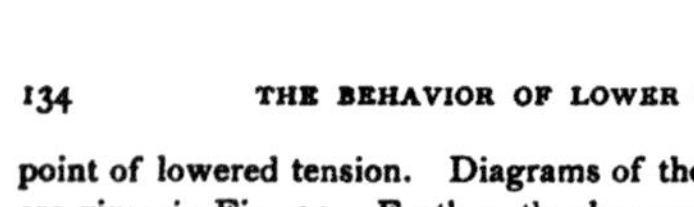

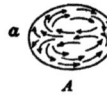

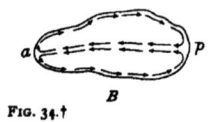

It is true that in the movements of *Amœba limax,* for example, one receives the impression of two sets of currents, one forward in the central axis, the other backward at the sides. But if the latter is studied carefully it is found that there is really no current here; the protoplasm is at rest, and the impression of a backward current at the sides is produced only by contrast with the forward axial current. *Amœba*

FIG. 34.

---

*All these facts are easily verified by placing a drop of clove oil on a slide in a mixture of two parts glycerine to one part 95 per cent alcohol under a cover supported by glass rods, as described in a previous paper by the present author (Jennings, 1902). By mixing some soot or India ink with the clove oil the currents are made evident.

† FIG. 34.—Currents in a drop of fluid when the surface tension is decreased on one side. *A,* the currents in a suspended drop, when the surface tension is decreased at *a*. After Berthold (1886). *B,* axial and surface currents in a drop of clove oil, in which the surface tension is decreased at the side *a*. The drop elongates and moves in the direction of *a,* so that an anterior (*a*) and a posterior (*p*) end are distinguishable.

limax contains usually a large number of fine granules, which in many cases extend to the very outer surface, so that it is not possible to distinguish an ectosarc, in the sense of a layer containing no granules. By watching the movements of these particles it is possible to determine the direction of the currents in the protoplasm. The movements in locomotion are usually as follows: At the anterior end there pushes forth from the interior a clear substance, which I will call the hyaloplasm. As this moves forward it spreads out laterally, till it reaches a position such that it forms a continuation forward of the remainder of the lateral boundary of the animal. Into this hyaloplasm flows then the granular endosarc. The granules flow forward, rapidly in the middle, usually more slowly near the sides. As it reaches the anterior end the central current spreads out in a fanlike manner, so that some of the granules approach closely the lateral borders of the Amœba (Fig. 35). They then stop, while the central part of the current passes on, following the advancing anterior end.

So long as one confines his attention to the Amœba alone, not observing external objects, one receives the impression that there are two sets of currents, an axial current forward, marginal currents backward. But as soon as one fixes his eye upon a particular granule in the apparent backward marginal current, and observes its relation to some external object, he discovers that no such current exists. The granule remains quiet, retaining continually its position with relation both to other granules in the edge of the Amœba and to objects external to the Amœba. Meanwhile the remainder of the substance of the Amœba is flowing past, so that the granule in question after a time comes to occupy a position at the middle of the length of the Amœba. At about this point it usually begins to move slowly forward again, though much less rapidly than the internal current. The nature of this slow forward movement we shall take up later (p. 166). The main portion of the body of the Amœba thus continues to pass the granule, and the latter finally reaches the posterior end. Here it usually remains quiet for a time (moving forward only as the posterior end is dragged forward). Then it is taken into the central current again, passes to the anterior end, and comes to rest as before, while the remainder of the Amœba passes it by; and this process is repeated

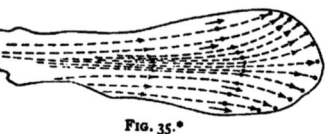

FIG. 35.

---

* FIG. 35.—Diagram of the movements of particles in an advancing Amœba. Each broken line represents the path of a particular particle.

indefinitely. In favorable cases I have repeatedly followed a single granule from the posterior end forward till it came to rest at the anterior end, then watched the body of the Amœba pass it by, until it was again at the posterior end and started forward anew. The course of a single granule is represented in Fig. 36. As is evident from this figure, the granule does not travel backward in any part of its course.

Not all the granules, however, remain quiet until they have passed to the posterior end. Many of them are taken again into the central stream before the entire body of the Amœba has passed them. Large granules usually stop only a short time, starting forward again before the middle of the Amœba has reached them; others are taken up at the middle or farther back, while many smaller granules reach the posterior end. But as a rule none show any movement backward, so far as I have observed.

It is not only at the margins of the Amœba, but also on the under surface, in contact with the substratum, that the ectosarc with its granules is at rest or moving slowly forward in the posterior half. This is evident when the lower surface of a transparent Amœba is brought into focus.

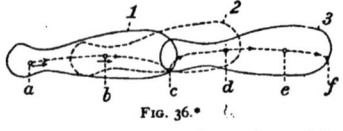

That excellent observer, Dr. Wallich, saw clearly many years ago that there is really no backward current, though at first view there appears to be such.

Fig. 36.*

It is only necessary to watch a specimen of Amœba carefully to become convinced that the appearance of a returning, as well as an advancing, stream of granules is illusory. The stream, it will be observed, is invariably in the direction of the preponderating pseudopodial projections. The particles simply flow along with the advancing rush of protoplasm. There is no return stream, but the semblance of one is engendered by one layer of particles remaining at rest whilst another is moving past them. (Wallich, 1863, *b*, p. 331.)

This statement of the facts my observations fully confirm.

In this account of the lack of backward movement in the granules of the ectosarc on the lower surface and at the margins I find myself

---

*Fig. 36.—Diagram of the movements of a single particle in Amœba, as seen from above. The particle begins at *a*, passes to *b* and then to *c*, at the anterior edge of the Amœba shown in the outline 1. The Amœba now passes forward to the position 2, and thence to 3, while the particle retains the position *c*; when the Amœba has reached the position 3 the particle is thus at its posterior end. Now the particle moves forward again, from *c* to *d*, and thence to *e* and *f*, thus coming again to the anterior edge. Here it stops, as at *c*, until the body of the Amœba has passed it. As the figure shows, the particle does not move backward in any part of its course.

at variance with certain statements of F. E. Schulze, Bütschli, and Rhumbler. I am aware that this conflict of my observations with those of the investigators named, who deservedly rank among the highest in the field at present under consideration as well as elsewhere, renders the utmost caution necessary in trusting to these results. Yet, with this consideration in mind, and with the confident expectation in undertaking the work that I should find the currents exactly as described by these authors, I have been unable to come to any result save that above set forth. The statements of Schulze (1875, pp. 344-348) deal with *Pelomyxa palustris* Greef. In this animal, according to Schulze, there are resting portions at the sides of the body, while from behind currents pass forward through the channel enclosed by these resting portions. At the anterior end the lateral parts of these currents turn outward and, finally, a little backward; any given portion passes backward but a short distance. The currents are shown by Schulze in a figure, a reduced copy of which is given herewith (Fig. 37). According to Schulze the currents have this form on the upper surface as well as at the sides—that is, a part of the current flows upward and backward on the upper surface. Bütschli (1892, Anhang, p. 220) confirms this account of the currents in Pelomyxa.

I regret that I have been unable to obtain specimens of Pelomyxa in order to examine these phenomena for myself. One would of course be bold to doubt the correctness of the observations of such investigators as Schulze and Bütschli, and it is possible that Pelomyxa differs from Amœba in this respect. Yet, as we shall see later (p. 149), the account given by these authors is certainly incorrect so far as the backward currents on the upper surface are concerned; it is possible, then, that the appearance of a backward current elsewhere was deceptive.

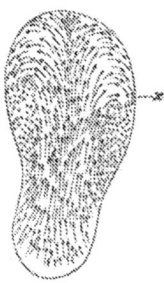

FIG. 37.*

Rhumbler (1898) describes the forward axial and the backward side currents in various species of Amœba, and considers such movements as typical, basing his theory of locomotion upon them. It seems probable that slight backward currents, such as were described by Schulze (Fig. 37), do occur at times at the sides of the advancing anterior end. The posterior part of the Amœba is narrow and rounded, the anterior part broad and thin. The current of endosarc flows from this narrow posterior portion into the broad anterior part

*FIG. 37.—Currents in a progressing Pelomyxa, as seen from above, after Schulze (1875). The longer arrows represent stronger currents.

and must therefore spread out; it would not be unnatural for the currents to flow even backward a little, as in Schulze's figure (Fig. 37), in order to fill the area just in front of the resting portion of the protoplasm (at x, Fig. 37). As we shall see, such movement is sometimes to be observed in inorganic fluids under similar conditions (p. 211). Whatever the explanation of the difference between my observations and those of the investigators named, the point of importance is that the backward current is not a constant nor an essential part of the locomotion of Amœba, so that it does not form a fitting basis for a theory of locomotion. Further, as we shall see, I am able to demonstrate conclusively the incorrectness of that conception of the nature of amœboid movement for which alone the account of the currents given by Bütschli and Rhumbler is significant.

It is evident that the method of movement here described is better adapted to the production of locomotion in a given direction than that which Bütschli and Rhumbler describe (see Figs. 30–33), since according to their account a portion of the substance of the body is first transported forward, then backward. In the locomotion as I observed it there is no such useless transportation of substance in a direction opposed to that in which the animal is traveling.

On the other hand, the movements as I have described them bear much less resemblance to those produced in drops of fluid by local changes in surface tension (Fig. 34). There is only the slight turning outward at the anterior end that can be at all compared to the backward flow of an outer layer in the inorganic drop. Rhumbler himself notes that in *Amœba angulata* there is often no such backward current to be seen (Rhumbler, 1898, p. 120), but bases his theory of the forward movement entirely on the cases where it (supposedly) does occur. In *Amœba angulata*, *A. verrucosa*, and *A. sphæronucleolus*, according to my observations, there is often no indication even of the turning out of the particles in a fanlike manner; they merely flow forward and stop for a time. Bütschli (1892, p. 199) notes that the backward current at the anterior end of Amœba, required by the surface tension theory, is very slight, but conceives it to be sufficient to fulfill the requirements of the theory.

MOVEMENTS OF UPPER AND LOWER SURFACES STUDIED EXPERIMENTALLY—ROLLING MOVEMENT IN AMŒBA.

Thus far we have left out of consideration the movement of substance on the upper surface of the Amœba. It is usually assumed that the condition here is the same as at the sides and on the under surface; thus Rhumbler gives a diagram, reproduced in my Fig. 33, *B*, showing the backward current of the upper surface. The positive observations on this point are those of Schulze (1875), Berthold (1886, p. 109), and Bütschli (1892, p. 220). These authors all agree that the backward current visible at the sides of the anterior end in Pelomyxa are clearly also present on the upper surface. It is not usually possible to observe particles moving backward on the upper surface of Amœba, nor even particles at rest, though this might be due to the fact that the granules have sunken downward, leaving the upper surface clear. But to decide whether the currents in Amœba are essentially like those produced in a drop of fluid by a local change in surface tension, it is most important to determine with certainty what is taking place on the upper surface.

Evidently the most natural way of doing this is to cause, if possible, some small object to rest upon or become attached to the upper surface of Amœba, then to observe the movement of this object. This can be done by mingling a considerable quantity of soot with the water in which the Amœbæ are found. Some of the soot particles settle on the

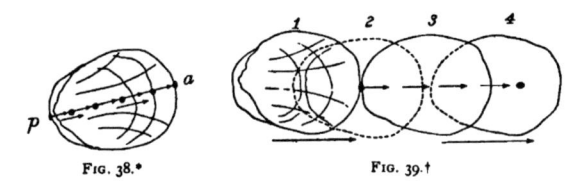

FIG. 38.* FIG. 39.†

upper surface of the Amœbæ, and in some species they adhere to this surface.

I was quite unprepared for the results of this experiment. *The upper surface of Amœba moves forward*, not backward, as required by the surface tension theory; nor is it at rest like the lower surface.

*FIG. 38.—Movements of a particle attached to the outer surface of *Amœba verrucosa*. When first seen the particle was at the posterior end (p); it then moved forward, as shown by the arrows, until it passed around the anterior end (a) to the under side. (The Amœba itself of course moved forward at the same time; no attempt is made to represent its movement in the figure.)

†FIG. 39.—Diagram of the movements of a particle attached to the outer surface of *Amœba verrucosa*, in relation to the movements of the animal. The Amœba is seen from above. In the position 1 the particle is at the anterior end of the Amœba. As the Amœba moves forward, it passes over the particle, which retains its place. Thus when the Amœba has reached the position 2 the particle is at the middle of its lower surface; when it reaches 3 the particle is at its posterior end. The particle then passes upward and forward, as shown by the arrows, so that when the Amœba reaches the position 4 the particle is in front of the middle, on the upper surface.

AMŒBA VERRUCOSA AND ITS RELATIVES.

In giving an account of the experiments which demonstrate this, I shall begin with species of Amœba in which pseudopodia are, as a rule, not formed, and the movements are uniform in character, since here the conditions are simplest from our present standpoint. For this purpose *Amœba verrucosa* Ehr., and particularly the transparent form known as *A. sphæronucleolus* Greef, are favorable. In these Amœbæ particles cling rather easily to the outer surface.

When a quantity of soot is added to the water containing Amœbæ of the species named, small masses cling to the surface of the animal. Such a mass, attached to the upper surface, shows the following movements: It passes slowly forward (Fig. 38), then over the anterior edge, and under the latter. Here it stops, while the Amœba continues to move forward. The mass of soot remains quiet until the entire Amœba has passed over it and it lies beneath the posterior end. It now passes upward again, to the upper surface (Figs. 39, 40), then forward once more to the anterior end. Here it goes under the Amœba

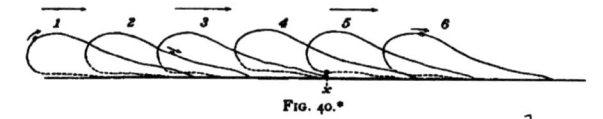

FIG. 40.*

as before, to be carried upward and forward again when the posterior end passes over it.

These observations are made with absolute ease, and there is no possibility of mistaking internal particles for external ones. Particles lying in the water outside the Amœba may be seen to become attached at the posterior end, to pass upward, lying distinctly outside the boundary of the protoplasm (Fig. 38, posterior end), then forward, till as they double the anterior end they are again seen sharply defined outside the boundary

---

* FIG. 40.—Diagram of the movements of a particle attached to the surface of *Amœba verrucosa*, in side view. In position 1 the particle is at the posterior end; as the Amœba progresses it moves forward, as shown at 2, and when the Amœba has reached the position 3 the particle is at its anterior edge, at *x*. Here it is rolled under and remains in position, so that when the Amœba has reached the position 4 the particle is still at *x*, at the middle of its lower surface. In the position 5 the particle is still in the same place, *x*, save that it is lifted upward a little as the posterior end of the Amœba becomes free from the substratum. Now as the Amœba passes forward the particle is carried to the upper surface, as shown at 6. (Thence it continues forward and again passes beneath the Amœba, etc.) The broken lines show that part of the surface of the Amœba which is at rest.

(Fig. 38, anterior end). Further, such particles, after making one or two revolutions, usually become detached and drop off.

It is thus clear that *Amœba verrucosa* and its relatives have what may be called a rolling motion; a given spot on the outer pellicula passes forward on the upper surface, downward at the anterior end, remains quiet on the lower surface, passes upward at the posterior end, and again forward. Its movement may be compared directly with the movement of a given point on the circumference of a wheel that is rolling forward. A diagram of the movement of a particle on the surface as it would appear in a side view is given in Fig. 40.

Certain details of the movements are interesting, and may best be brought out by description of specific observations. In one case two small particles had become attached, a short distance apart, to the surface of a specimen of *Amœba sphæronucleolus*. They were at first side by side and a little to the right of the middle line, one somewhat farther to the right than the other (Fig. 41). They moved forward in parallel courses, and reached the anterior edge at the same time, passing over the edge and to the under surface. It now required two and one-half minutes for the Amœba to pass over them, during which time they remained nearly or quite at rest. They then moved upward to the upper surface and forward again. The one nearer the middle line moved a little faster than the other, reaching the anterior edge in two and three-quarter minutes, while the lateral one required three minutes. Both emerged at the posterior end again at the same time, the central one having remained quiet three and one-fourth minutes, while the lateral one had been three minutes at rest. The next forward course required, respectively, but one and one-half and two minutes, the central particle moving the more rapidly. The two particles were no longer side by side, the central one being now a little in advance. The latter spent after the next turn two and one-half minutes on the under surface, while the lateral particle spent but two minutes, so that they came up from the posterior end again at the same time.

FIG. 41.*

The two particles started forward again and had reached the middle of the upper surface when the Amœba ceased its forward movement, loosened its anterior end from the bottom, and became attached by its posterior end. After five minutes it began to move again, but now in

---

* FIG. 41.—Paths of two particles attached to the outer surface of *Amœba sphæronucleolus* as described in the text. That portion of the paths which is on the lower surface is represented by broken lines. (No attempt is made to represent the forward movement of the Amœba in this figure.)

the opposite direction, so that the former posterior end became anterior. At the same time the two particles reversed their former motion and began to travel back in the direction from which they had come—that is, toward the new anterior end. They were observed to make several complete turns about the Amœba while moving in this new direction. I will not, however, add further details, as those above recounted are sufficient to give a conception of the main features of the movement.

Thus two definite points on the surface of an Amœba may retain nearly the same relation to one another for five or six complete revolutions, though their distance apart and their relative position may vary a little. The reversal of the direction of rotation when the direction of locomotion is reversed, described in the above case, I have seen many times.

The direction of the movement of particles on the outer surface is the same as that of the underlying particles of endosarc. The rate is also about the same as for the endosarc, though often, or perhaps usually, the outer particles move a little more slowly than those in the endosarc.

It is not merely a thin outer layer that has the rolling movement. This is demonstrated by the movements of bodies that are partly embedded in the substance of the Amœba. For example, a large Euglena cyst had become attached to the hinder end of an *Amœba sphæronucleolus*. The cyst was carried upward and forward on the upper surface, and at the same time it began to sink into the protoplasm, so that when it had reached the anterior edge it was partially embedded. It was then rolled under, remained at rest on the under surface in the usual way, and came up at the posterior end. It was now deeply sunk in the protoplasm, yet it moved forward in the usual way. By the time it had reached the anterior edge again it no longer protruded above the surface at all. After turning the anterior edge again it sank completely into the body, still surrounded by a layer of ectosarc, so that it passed to the interior of the Amœba as a food body. I have repeatedly seen bodies which were thus carried forward on the upper surface gradually taken in as food. They always continue the forward movement even when completely embedded in the ectosarc. It is thus evident that the whole thickness of the ectosarc partakes of the forward movement. The forward stream in ectosarc and endosarc are one and continuous.

The relation of the movements of the outer layer to the lines and wrinkles seen on the upper surface of *Amœba verrucosa* and its relatives is of interest. There are usually two sets of these wrinkles, one set diverging from the posterior end toward the direction in which the animal is moving, the other set forming a number of curved lines parallel to the advancing edge (Fig. 38). These wrinkles and the areas which they enclose do not change markedly as the Amœba advances, so that the outer surface of the body seems to be quite at rest. It is this fact, I believe, that has prevented the true nature of the movement in these species from being recognized before. Thus Penard (1902, p. 118), after a thorough study, accurate so far as it goes, of the movements of *Amœba verrucosa*, notes that many facts point to the existence of a permanent contractile outer layer, but holds that the permanence of certain lines and patterns on the upper surface in a moving Amœba is crucial against the idea of a rolling movement such as I have shown above to actually occur. In reality these wrinkles are not static structures, but dynamic, *i. e.*, the substance of which they are formed is in continual motion; they are like the permanent ripples on the surface of a stream where the latter crosses an obstruction. The wrinkles indicate the direction of movement of the substance, the longitudinal wrinkles being parallel to the lines of motion, the others transverse to them. A particle on the upper surface may move parallel with the longitudinal wrinkles, at a constant distance from them, or it may move directly along one of these wrinkles, for the whole length of the latter. On coming to one of the transverse wrinkles the particle moves over it with a sort of jerk, as if it had passed over a ridge or step, as indeed it has. Thus the lines and the areas enclosed by them remain constant, while the substance of which they are composed moves onward.

When the Amœba changes in a marked degree its direction of movement, so as to follow, for example, a course at right angles to the previous one, the wrinkles on the surface usually slowly disappear, then after the movement has become well established in the new direction, new wrinkles appear in correspondence with the movement.

When such a change of course occurs, any particles on the upper surface, which were moving toward the anterior edge, change their course in correspondence with the new direction of progression. Fig. 42 represents a case of this kind, where an *Amœba verrucosa* bore on its upper surface a minute particle of débris (*a*) and a spherical cyst of Euglena (*b*). Both moved forward over the stretch *x-y* (Fig. 42, *A*). Now a little methyl green (*m*) was allowed to diffuse against the left side of the Amœba. The animal changed its course, moving to the right. At the same time the two objects *a* and *b* changed their direction of movement, traversing the stretch *y-z* (Fig. 42, *B*) until they reached the new anterior edge of the Amœba, and were carried underneath.

The free-moving (upper) surface and the resting (lower) one in contact with the substratum may exchange rôles at any time when the contact with the substratum is changed. Thus, a specimen was creeping on the slide and bearing on its upper surface a small granule, which was moving forward in the usual way. The Amœba stopped and raised its anterior edge, which came in contact with the cover glass; it then loosened itself entirely from the slide, while its upper surface became attached to the cover. It now began to move forward on the cover glass. The granule on the upper surface now remained quiet, until it was reached by the posterior end, when it passed downward to the lower free surface, there moving forward in the usual way. Upper and lower surfaces had completely exchanged rôles. In a similar way I have seen the thin lateral edge of a specimen become the middle of the upper moving surface.

Objects of the most varied sort cling to the surface of *Amœba verrucosa* and its relatives. I have seen the following attached to the surface and showing the typical movements: Particles and masses of soot, granules of India ink, motionless bacteria, diatom shells, dead flagellates, masses of débris, cysts of Euglena, a small *Amœba proteus* (the latter was inclosed after it had passed to the under surface). Usually only one or two small objects are seen attached to any given specimen, but to this extent the phenomenon is very common, so that it seems rather surprising that the movements of such particles should not have been described before.

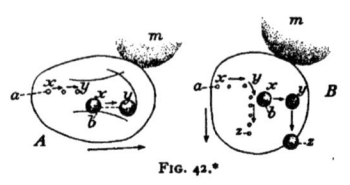

FIG. 42.*

A number of other points must be set forth before we can form a clear conception of the movements of these Amœbæ. The species under consideration are much flattened and have usually an oval form as they move forward, the anterior-posterior axis being the longer, while the posterior end is the more pointed (Fig. 38). Not the whole lower surface is in contact with the substratum, but only a band at the anterior and lateral margins. In an Amœba that was creeping on the

* FIG. 42.—Movement of bodies attached to the surface in *Amœba verrucosa*, when the direction of locomotion is changed. *a*, A small granule; *b*, a Euglena cyst. In *A* the Amœba is progressing to the right, as shown by the large arrow; the two bodies attached to the surface moved in the same direction, traversing the stretch *x–y*, as shown by the small arrows. At this point a solution of methyl green (*m*) was allowed to diffuse against the surface; the Amœba thereupon changed its course, as indicated by the large arrow of *B*. At the same time the bodies *a* and *b* changed their course, traversing the stretch *y–z*. The stretch *x–y–z* in *B* shows thus the path of the attached bodies before and after the reaction.

lower surface of the cover glass I was able to define with some accuracy the parts that were attached and those that were not. A small flagellate was moving briskly about between the Amœba and the cover glass, but its excursions were limited by a visible line running parallel with the anterior edge of the Amœba and extending at the sides back to about one-third the animal's length from the rear (Fig. 43, *a-a-a*). The zone between this and the margin was pressed close to the glass, and was evidently attached to it. The more pointed posterior end was held quite away from the glass, leaving a broad passageway through which the flagellate finally escaped.

The results of this observation were confirmed by another. An *Amœba verrucosa* in full career was suddenly turned on one lateral edge by a strong current from a rotifer, and its upper edge coming in contact with the cover glass, it remained in that position some time without change of form. It could be seen that the under surface was concave, the edges very thin and flat, while the posterior portion was thick and arched (Fig. 44).

It is clearly at the advancing edge of the animal that the most active movements are taking place. Here the hyaloplasm may be seen to push forward in a series of short waves, the anterior edge of each becoming attached to the substratum. At the same time, of course, an equivalent amount of protoplasm becomes detached from the substratum along the line *a-a-a*, Fig. 43, though this does not take place in waves, so far as observable. The anterior wave must in some way pull upon the upper surface of the Amœba, bringing it forward, and dragging with it the elevated sac-like posterior end. A certain feature of the advance of the anterior edge seems of much significance. Each wave seems to arise just behind the previous anterior boundary line and overlaps it, leaving it buried. This line often remains visible for a short time after the new wave has been formed. The new wave rolls over the preceding one in such a way that its original upper surface becomes applied to the substratum. This is demonstrated by the rolling under of small objects on the upper surface of the advancing wave. A diagram of the movement at the anterior edge is given in Fig. 45. The movement can be imitated roughly by making a cylinder of cloth, laying it flat on a plane surface, and pulling forward

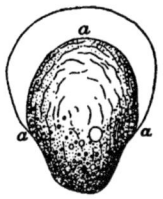

FIG. 43.*

* FIG. 43.—Attached surface of *Amœba verrucosa*, creeping on the lower surface of the cover glass. The unshaded portion in front of the line *a-a-a* is attached to the substratum, while the shaded portion is free and raised slightly above the substratum.

the anterior edge in a series of waves. The entire cylinder then rolls forward just as the Amœba does.

The essential features of the movement seem to be (1) the advance of the wave from the upper surface at the anterior edge ; (2) the pull exercised by this wave on the remainder of the upper surface of the body, bringing it forward. Most of the other phenomena follow as consequences of these two. The flowing forward of the granules of the endosarc seems to demand no special explanation, since a fluid containing granules within a rolling sac must necessarily flow forward as the sac rolls. By the movement forward of the anterior end a space is left free ; by the rolling forward of the posterior end the fluid is piled up and pressed upon, and must flow forward into the empty space in front. Possibly there may be other causes at work in producing the endosarcal currents, but such currents would be produced without other cause in a sac moving as Amœba does.

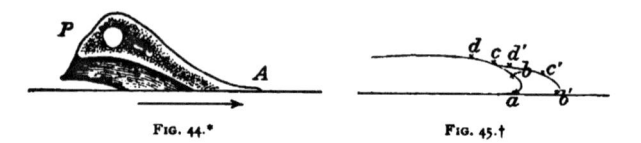

FIG. 44.*   FIG. 45.†

OTHER SPECIES OF AMŒBA.

Thus far we have dealt only with Amœbæ of rather constant form, which do not produce pseudopodia, or only rarely do so. We must now take up species in which the form is changeable and the movements varied. Of such species I have studied chiefly *Amœba limax*, *A. proteus*, and a smaller Amœba, which I take to be *Amœba angulata* Meresch. In these species the outer surface is not viscid, except at the posterior end, so that small objects rarely cling to it. It is, therefore, much more difficult to determine the direction of movement of the upper surface than in *Amœba verrucosa* and its relatives. Yet, by mixing soot with the water, and devoting a sufficient amount of time and patience to the work, one can obtain as many observations as he desires. The soot settles upon the upper surface in particles or masses

---

* FIG. 44.—Side view (partly an optical section) of a creeping *Amœba verrucosa*, showing the thin anterior edge (*A*) attached to the substratum, and the high posterior portion (*P*) with a cavity beneath it.

† FIG. 45.—Diagram of the movement at the anterior edge of *Amœba verrucosa*. The region $b-c$ pushes out, taking up the position $b'-c'$, and pulling forward the region $c-d$, so that it comes to occupy the position $c'-d'$. The point a remains in its place.

and its movements can be followed; at times, also, objects actually cling to the surface, as in the other species.

The results are essentially the same as in the species already described; foreign particles resting upon or clinging to the upper surface are carried forward to the anterior edge. Here they roll over the edge, passing beneath the Amœba, which now moves across them. As a rule in these species particles do not cling to the surface after passing to the lower side, so that they are left behind when the posterior end passes over them. Sometimes they do thus cling, however, and in such cases I have seen them pass upward at the posterior end and again forward, exactly as in *A. verrucosa* and its relatives. In order that my statements may not remain abstract and general, I copy a few observations from my notebook, all relating to *Amœba proteus*.

1. A large particle of débris with bits of soot attached to it was seen lying on the upper surface just behind the middle. It was carried forward to the anterior end and over the edge. Then it came to rest on the bottom, and the Amœba crept over it till it was passed by the posterior end and left behind.

2. A number of soot particles on the upper surface just in front of the middle were carried forward, changing their direction as the protoplasmic currents beneath them changed direction. They were finally carried over the anterior edge.

3. A small mass of soot was lying on the middle of the upper surface. It moved forward in the same way as the endosarcal granules underneath. The latter changed their direction of movement several times; the soot mass changed correspondingly at the same time. It was finally carried over the anterior edge, where it could be seen clearly separate from the Amœba.

4. A large mass of soot one-quarter the size of the Amœba was carried forward on the upper surface for a distance, but fell off at the side before reaching the anterior end.

5. Two small masses of soot lying on the upper surface of the posterior end were carried forward over the anterior edge.

6. Several small particles were clinging to the lower surface of the posterior end. They passed upward, one of them around the very middle of the posterior end, to the upper surface; here they were carried forward and over the anterior edge.

I could add a large number of such observations.

On the under surface the particles are quiet, as I have shown before (p. 136). At the lateral margins the edges of this quiet lower surface are seen, so that particles situated here are usually likewise quiet, until they have reached the posterior part of the Amœba (see p. 135).

As to the details of the movement of the upper surface, the following are important. Particles situated on the upper surface move usually at the same, or nearly the same, rate as the granules beneath them, in the endosarc. The movement of the surface particles follows exactly that of the endosarc beneath them, changing in direction when the latter changes. Two particles close together on the upper surface may thus diverge or even flow in opposite directions, carried by two different currents which are visible in the endosarc. Any portion of the ectosarc, like any portion of the interior, may stop at any time, while other parts flow onward. One may thus see at times a particle at rest on the upper surface of a moving Amoeba. Isolated observations of this kind might lead one to suppose that the upper surface, like the lower, remains at rest while the endosarc passes forward. But when a particle on the surface is at rest, one will usually find, by a proper change of focus, that the endosarc beneath it is likewise at rest. It is, of course, well known that certain portions of the endosarc may be at rest while the remainder is in movement (see Rhumbler, 1898, p. 122). In the same way a portion of the outer layer may sometimes be at rest while the adjacent endosarc is in motion; this, however, is rather unusual.

We may sum up our results thus far in the following statements: *In an advancing Amoeba substance flows forward on the upper surface, rolls over at the anterior edge, coming in contact with the substratum, then remains quiet until the body of the Amoeba has passed over it. It then moves upward at the posterior end, and forward again on the upper surface, continuing in rotation as long as the Amoeba continues to progress. The motion of the upper surface is congruent with that of the endosarc, the two forming a single stream.*

HISTORICAL, ON ROLLING MOVEMENTS IN AMŒBA.

The possibility that Amoeba progresses by a rolling movement was discussed by Claparède & Lachmann (1858). In *Amœba limax* and *Amœba quadrilineata* (= *A. verrucosa*), according to these authors, the general appearance of locomotion is in many respects in favor of this view: "On croit positivement voir l'animal rouler sur lui-même" (p. 435). But this correct view is rejected (in the text) because of the (supposed) permanence in the position of the contractile vacuole. Claparède & Lachmann insist that the contractile vacuole is situated in the ectosarc; hence, they argue, if there were a rolling movement of the ectosarc, the vacuole would necessarily partake of the movement! In a note on p. 437 it is stated, however, that Lachmann personally believed the motion to be of this rolling character. "Il croit s'être assuré que *l'A. quadrilineata* roule sur elle-même." According to Claparède & Lachmann, Perty held this view also.

Dr. Wallich shared the correct opinion of Lachmann and Perty. This excellent observer unfortunately often gave his results in the form of mere brief general statements, so that one cannot judge how much evidence he had for them, and little attention has, therefore, been paid them. But it is singular how many of these statements show themselves to be correct, even in opposition to later work. Concerning the matter in question, Wallich has the following:

> In short the effect is similar to that which would be produced were an empty and transparent bladder or caoutchouc sac, containing granular bodies of greater specific gravity than the viscid fluid within which they were sustained, to be rolled along a plain surface. (Wallich, 1863, *b*, p. 331.)

He makes no attempt to demonstrate the truth of this correct comparison, and does not develop the matter beyond the mere statement given above.

Schulze (1875), Berthold (1886), and Bütschli (1892), as we have seen, agree in stating that the currents on the upper surface at the anterior end in Pelomyxa are backward. In view of the great authority of these writers we should be compelled to suppose that the movement in this animal is of an entirely different character from that found in the various species of Amœba, but for a most fortunate circumstance. The only previous demonstrated observation of the forward movement of the upper surface in the Rhizopoda relates precisely to Pelomyxa, and was made by an investigator closely associated with Bütschli. It was not until my work was finished and the present paper written that I came across the note of Blochmann (1894) on the movements of Pelomyxa. Blochmann shows that the movement of substance on the upper surface of Pelomyxa is forward, just as we have found it to be in Amœba. The outer surface of Pelomyxa is covered with fine cilia-like projections. By observing these projections Blochmann had no difficulty in seeing that they move forward on the upper surface. The rate of movement was the same as that of the internal forward current. Bütschli (1892, Appendix, p. 220) had already observed, greatly to his surprise, that there is a forward current in the water next to the surface of an advancing Pelomyxa, this current being exactly the reverse of that called for by the theory that the motion is due to a lowering of the surface tension at the anterior end.

Bütschli (*l. c.*) attempted to save the surface tension theory by suggesting that it was only a thin outer layer that moves forward. The currents in the moving animal would then be as follows: A forward current within, a backward current just beneath the surface, a forward current in a thin layer on the surface. It is possible that this complicated arrangement of currents might be brought into harmony in some way with the surface-tension theory of the movement, though it is rather difficult to see how the forward movement of the outer layer would be produced. As we have seen, Blochmann found that the internal and external forward currents move at the same rate and in the same direction. It is difficult to explain how this should occur if the two are separated by a layer moving in the opposite direction. But Blochmann accepted Bütschli's suggestion, and attempted to give some evidence in its favor. He says that one sees the outer current, with the movement of the projections, in places where the marginal current has come to rest, and that the outer and internal currents then move at the same rate, separated by the resting marginal layer. Now one can receive exactly this impression in Amœba in the following manner: The margins where they are pressed against the substratum are at rest. Just above this region, and often visible in the same focus, there is the forward current, which is visible on the one hand on the surface (through the movements of the projections); on the other hand, in the interior (through the movements of the granules of the endosarc). Unless one is on his guard as to the slight difference in level, one might seem to see two currents separated by a resting layer, particularly if the probability that this were true had been suggested beforehand. It is notable that Blochmann describes nowhere outer and inner forward currents, separated by a marginal backward current, as would be required by the modified surface-tension theory.

We have demonstrated above, for Amœba at least, that the forward movement is not confined to a thin outer layer, but extends from the outer surface to the endosarc (p. 142); in other words, that the outer surface moves in continuity with the internal substance.

Rhumbler (1898, pp. 126–130) discussed at length the possibility of explaining the movement of Amœba by means of a rolling sac of ectoplasm, only to come to the conclusion that it was impossible. Rhumbler's discussion of this matter is an excellent example of the fact that acumen and excellent reasoning may lead one astray in scientific matters when the observational basis for the reasoning is not secure. What chiefly misled him was an incorrect idea as to the direction of the currents in the substance of Amœba, particularly his assumption that there is a backward current on the upper surface. The diagram which he gives of the currents as they must occur in an Amœba moving in a rolling manner (Fig. 33, *A*, p. 133) is, therefore, much more nearly correct than that in which he shows what he considers the really existing currents (Fig. 33, *B*).

Rhumbler's conception as to the necessary movements in the substance of an Amœba progressing by rotation is, however, incorrect in one particular, so that his diagram (Fig. 33, *A*) does not correspond to the facts in this point. He assumes that there must be a backward current on the lower surface, as indicated by the lower (heavier) arrows in his diagram (Fig. 33, *A*). This backward current does not exist, and is theoretically unnecessary, as may be seen by making a cylinder of cloth and moving it in the manner described above (p. 145). The under surface remains at rest until it passes upward at the posterior end (*cf.* Fig. 40). Rhumbler held that this backward current below, with the forward current above (Fig. 33, *A*), must set the endosarc in rotation; "the endoplasma granules would themselves necessarily all move, like the particles of the ectosarc, in circular or elliptical courses" (p. 128). The absence of such circular or elliptical paths for the granules of the endosarc would then speak against the method of movement by rotation of the ectoplasm. But since there is no such backward current as Rhumbler assumes, and not even the particles of the ectosarc move in circular or elliptical courses, this objection falls to the ground.

Further, Rhumbler seems to assume that for locomotion by a rotary movement of the ectosarc, the latter must necessarily be a "sharply defined persistent organ," and that its contractions could only be due to preformed, permanent fibers, in a definite arrangement. Rhumbler is able to show of course that these two assumptions are probably incorrect, and considers that this weighs against the possibility of movement in the manner characterized. But both these assumptions are unnecessary. The rotation demonstrably does occur, yet the permanent, sharply defined ectosarc with definitely arranged persistent fibers does not exist, as Rhumbler has set forth, and as must be evident to anyone who studies for a long time the changes of form and movement in Amœba. As we shall see later, a simple drop of fluid, with no differentiated outer layer, may move in the same manner.

Penard (1902) also discusses the possibility of movement by rotation of the ectosarc in *Amœba verrucosa* (= *A. terricola*). His study of the movements is excellent and he gives as a possibility on p. 115 what is really in its main features a nearly accurate statement of the method in which locomotion actually occurs, only to reject this possibility later. The ground for this rejection is as follows: The posterior end of the Amœba often bears an irregular saclike projection (what Penard calls the "houppe"); this may be much wrinkled or covered with projections. This wrinkled sac retains its position; in *Amœba verrucosa* it is covered, like the rest of the body, with a resistant cuticula, which can be dissolved only with great difficulty and very slowly.

If the Amœba rolled on itself in progressing, the posterior part of this membrane would necessarily follow the movement and pass little by little forward, which is contrary to the facts. The best manner of assuring one's self of the immobility of the pellicle is to look very attentively at the surface of the contractile vacuole; there one sees almost always very fine folds, forming angles and varied patterns; these angles and these patterns remain for a long time absolutely the same, which shows that nothing has changed place. (Penard, 1902, p. 118.)

In all the specimens of *Amœba verrucosa* and *A. sphæronucleolus* in which I have studied the matter, the posterior part of the outer membrane does follow the movement. Particles clinging to the outer surface of the hinder part of the ectosarc pass upward over the wrinkled saclike posterior end and forward on the upper surface. In so doing they pass directly across the wrinkles on the body surface, as set forth on p. 143. Had Penard chanced to see the movements of a particle attached to the outer surface of the body he could not have been misled by the apparent permanence of the surface wrinkles.

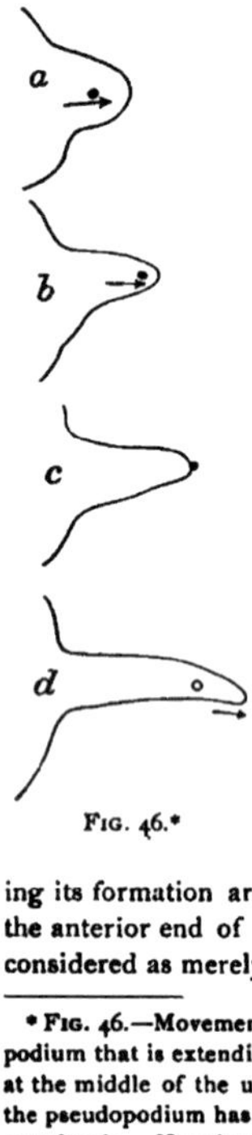

FORMATION AND RETRACTION OF PSEUDOPODIA.

Thus far the phenomena in *Amœba proteus* and its relatives are essentially like those found in *Amœba verrucosa*. At times *Amœba proteus* flows forward as a single simple mass; then its locomotion may be compared directly in its chief features to that of *Amœba verrucosa*. But in *Amœba proteus* and its relatives the movement is, of course, usually much complicated by the formation of pseudopodia. In considering the way in which these are formed we must deal separately with two different cases, depending on whether the pseudopodium when sent out is or is not in contact with the substratum.

SURFACE CURRENTS IN THE FORMATION OF PSEUDOPODIA IN CONTACT WITH THE SUBSTRATUM.

FIG. 46.* When the pseudopodium is sent out in contact with the substratum, the phenomena accompanying its formation are essentially the same as those which take place at the anterior end of an advancing Amœba; the latter may indeed be considered as merely a large pseudopodium. Even when the pseudo-

* FIG. 46.—Movement of a particle attached to the outer surface of a pseudopodium that is extending in contact with the substratum. At *a* the particle is at the middle of the upper surface; at *b* it has nearly reached the tip. When the pseudopodium has reached the length shown at *c* the particle has passed over its tip. Here it remains, so that at *d*, when the pseudopodium has become longer, the particle is still at the same level, but on the under surface of the pseudopodium, some distance behind the tip.

podia are slender and pointed, the protoplasm flows forward (toward the point) on the free upper surface and in the interior, while on the side which is in contact with the substratum the protoplasm is at rest. I have often seen small particles which had been brought forward on the surface of the Amœba carried out to the tip of a pseudopodium on its upper surface, finally rolling over the point and becoming covered by the advancing protoplasm (Fig. 46).

FORMATION OF FREE PSEUDOPODIA.

When a pseudopodium is sent out directly into the water, so that its surface is free on all sides, it is much more difficult to determine the nature of the movement. Particles rarely cling to the surface of such a pseudopodium, and without this aid one cannot be certain what the movement of the outer layer is. However, by devoting several entire days under most favorable conditions to the determination of this point, I collected a number of observations which demonstrate clearly the nature of the movement. The point of special interest is, here as elsewhere, whether there is a backward current on the surface of the advancing pseudopodium, as represented in the diagram from Rhumbler, Fig. 31. To this the observations give a negative answer. Particles clinging to the surface of a pseudopodium, whether at the tip or at the base or at any intermediate point, are uniformly carried outward, in the same direction as the tip. Particles situated at a certain distance from the tip of a short pseudopodium maintain the same distance as a rule when the pseudopodium is lengthened, though in so doing they are carried far out from the body. Sometimes the tip moves outward a little faster than the parts behind it, the pseudopodium thus becoming more slender as it extends, but all parts agree in being carried outward. A number of examples of actual observations will make this point clear.

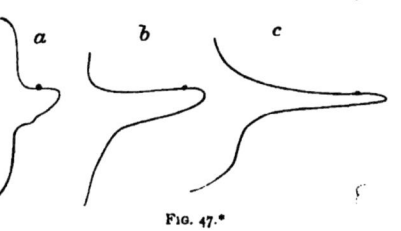

1. *Amœba angulata:* When first observed there was a short pseudopodium in front, projecting freely into the water. A small particle was attached to the surface at about the middle of its length (Fig. 47, *a*).

---

* FIG. 47.—Successive stages in the formation of a free pseudopodium, showing the movement of a particle attached to its surface. The particle moves outward, keeping at approximately the same distance from the tip.

The pseudopodium lengthened, carrying the particle with it, the latter maintaining its distance from the tip nearly or quite constant, but being carried far from the body (Fig. 47, b). The pseudopodium finally became very long and slender (c), the particle remaining attached near the tip.

2. *Amœba proteus:* A particle clinging to the surface of one side, near the anterior end. A pseudopodium was formed at exactly this point, extending freely into the water, so that the particle was borne on the tip of the pseudopodium. It maintained this position while the pseudopodium was extending, and was still found at the tip after the pseudopodium had become long and slender (Fig. 48).

The third example which I give is one of much interest, because it shows the movements of a given point on the surface in both the retraction and extension of pseudopodia, as well as in transference from the posterior to the anterior region of the body.

3. *Amœba proteus:* When first observed the animal was rather slender, creeping in a certain direction, and with two long pseudopodia at the posterior end, extending, one on each side, at right angles to the axis of progression (Fig. 49, a). The left pseudopodium was the longer, and bore at about one-fourth its length from its base a small particle (*x*) attached to its surface by a very short stalk in such a way that it was seen in profile (Fig. 49, a). The pseudopodium was not in contact with the bottom, and was slowly retracting, its internal contents flowing into the body, while the pseudopodium itself shortened. As this occurred the particle approached the body and finally passed on to its surface (*b, c*) while the pseudopodium was yet of considerable length. It was evident that the shortening of the pseudopodium took place chiefly at its base, since the part between the base and the particle *x* had become incorporated with the body when the portion between *x* and the tip had changed only a little in length (*b, c*). This distal portion apparently did become somewhat shorter at the same time, while its surface became slightly wrinkled. By the time the tip of the pseudopodium had united with the body (*d*) the particle *x* had moved a considerable distance forward on the latter. The posterior portion of the body was here thick, and the particle was still seen in profile, though it was some distance above the substratum. It now moved

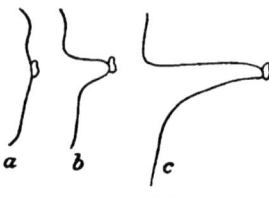

FIG. 48.*

---

* FIG. 48.—Movement of a particle attached to surface of an Amœba at point where a free pseudopodium is pushed forth. The particle remains at the tip.

forward very slowly (*c, d, e*) till at *f* it passed to the upper surface. It then moved rapidly forward, occupying successively the positions indicated by the line of circles in *g*. (The Amœba itself was, of course, progressing; no attempt is made in the diagram to represent its change of position.) Finally the particle *x* had nearly reached the anterior end, when the latter forked, sending two pseudopodia upward and forward into the water (*g, h*). The particle *x* was at first at the base of the right pseudopodium. This was now projected forward as a very long, slender pseudopodium bearing the particle *x*. The latter was carried steadily out from the body, maintaining almost exactly its original distance from the tip of the pseudopodium (*h, i, j*). It is

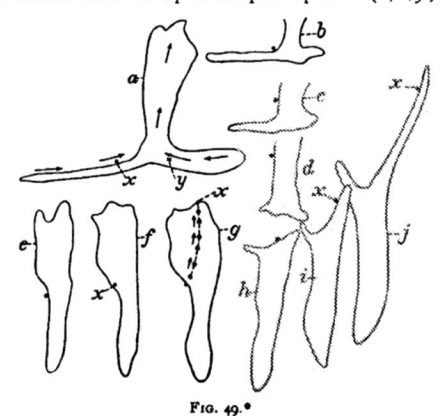

FIG. 49.*

possible that as the tip became very slender its distance from *x* became slightly greater as if, by a circular contraction of the intervening part, the tip were forced further out; but there was no movement backward of *x*; on the contrary, it moved steadily forward, its distance from the base of the pseudopodium continually increasing. Unfortunately at this point the animal passed under a mass of débris, so that I was unable to trace further the history of that point on the body surface marked by the particle *x*.

I have, altogether, about a dozen observations showing this outward movement of particles on the surface of free pseudopodia. The three

---

* FIG 49.—Movements of a particle (*x*) attached to the surface of Amœba in passing from a pseudopodium at the posterior end over the body to a pseudopodium at the anterior end. For explanation see text.

examples above given are typical for all. They show the following as to the manner in which the pseudopodia are formed when they are projected freely into the water.

1. The pseudopodium grows in length chiefly from the base, so that any part on the surface retains nearly its original distance from the tip.

2. The increase in surface as the pseudopodium grows is not produced by the flowing outward and backward of the endosarc at the tip with its transformation into ectosarc (as represented by Fig. 31), but by the transference of a portion of the surface layer of the body to the pseudopodium. The same substance remains at the tip of the pseudopodium from the beginning (observation 2, p. 154; I have other observations showing the same thing).

3. Thus the movement of the free pseudopodium is like that of the pseudopodium in contact with a surface, save that in the latter case one side is held back by attachment to the substratum. In the free pseudopodium all sides move outward; in the attached one, all sides but one.

The outer layer of the body in its transference to the pseudopodium may doubtless become thicker or thinner or be otherwise modified. As will be shown later, I am not at all inclined to deny the possibility of the transformation of endosarc into ectosarc, and *vice versa*. The observations show, however, that this transformation of substance does not, as a rule, take place in pseudopodia by means of the "fountain currents" represented in the diagrams from Rhumbler (Figs. 30–32).

Further, the surface of the pseudopodium may be increased by the flowing into it of the endosarc, producing a sort of stretching of the outer layer, involving, of course, the appearance at the surface of portions of substance which were before covered.

WITHDRAWAL OF PSEUDOPODIA.

In the withdrawal of pseudopodia the process is the reverse of that occurring in the formation of pseudopodia, as is shown in case 3, above (p. 154, Fig. 49). The basal parts of the pseudopodial surface first pass on to the body, followed by the distal portions.

The withdrawal of pseudopodia shows certain other features that are of importance for the understanding of the mechanism of amœboid movement. The process differs somewhat in different cases, depending on whether the withdrawal is slow or rapid. When the pseudopodium is slowly withdrawn, its surface may remain perfectly smooth, the decrease in surface keeping pace with the decrease in volume, until the pseudopodium has quite disappeared. But when the withdrawal is more rapid the surface becomes thrown into folds or warty prominences of various sorts. This is more common than retraction without the formation of such prominences. *Evidently the volume decreases so fast that the decrease in surface cannot keep pace with it*, so that the surface is thrown into folds. This phenomenon is particularly interesting in its bearing on the theory that would account for the retraction of pseudopodia by the action of surface tension. On this theory we should naturally expect the surface to remain smooth, and by no means to be thrown into folds, since it is by the tendency of the surface to decrease that the decrease in volume is accounted for; the decrease in volume should not, therefore, precede the decrease in surface. This matter will be taken up later.

As the pseudopodium decreases in size, the fluid endosarc, of course, flows out of it and joins the endosarc of the body. The backward current begins at the mouth or inner end of the pseudopodium, and gradually extends backward to near the tip; the current is most rapid in the central axis of the pseudopodium, and in this axis it is most rapid at the inner end.*

Where is the impelling force in the outflow of the endosarc and the decrease in size of the pseudopodium? The observations seem to suggest several factors here. The fact that the ectosarc of the pseudopodium passes on to the body when the pseudopodium shortens, as is shown in Fig. 49, *a, b, c,* indicates that the ectosarc of the body exercises a pull on the outer layer of the pseudopodium, drawing it inward. This would, of course, force the fluid endosarc into the body. But this would not account for the wrinkling and roughening of the outer surface of the pseudopodium, which is so prominent a feature in the withdrawal. For this there are two conceivable causes. (1) The ectosarc itself may contract actively, driving out the endosarc. If the real contractile portion of the ectosarc is not on the outer surface (in the cuticula, as it has sometimes been called), but in a deeper layer, then the outer surface would be thrown into folds or prominences as contraction occurs. (2) On the other hand, it is conceivable that the endosarc might be drawn out of the pseudopodium, the latter collapsing and becoming wrinkled as a result. This is the explanation given by Bütschli (1892, p. 201). This view would have to assume some force pulling on the endosarc at the mouth of the pseudopodium, and sufficient viscosity in the endosarc so that a pull thus exercised would draw out the whole mass contained within the pseudopodium. Thus, in Fig. 49, *a,* the general advancing current within the body of the Amœba might be thought to exercise a pull at the point *y* in the direction of the arrow; if the endosarc were

---
*This account differs from that given by Bütschli (1880, p. 116), according to whom the withdrawal of the pseudopodium begins at the tip. The observations present no difficulty, and I am unable to understand how Bütschli came to this result. In a large pseudopodium the method of retraction described above is evident.

sufficiently viscous the entire mass of endosarc would be withdrawn, and the pseudopodium would collapse.

There are certain facts that speak against this second view. Thus, the endosarc often passes out when there is no current away from the mouth of the pseudopodium, so that there can be nothing pulling upon the endosarc. A pseudopodium may be withdrawn when the animal is otherwise quiet; or, when the animal is stimulated strongly, all the pseudopodia may be withdrawn at the same time, while there is no endosarcal current in the body of the animal. A striking case that belongs here is sometimes to be observed in *Amœba radiosa*. This animal frequently floats in the water, with many long, pointed pseudopodia radiating in all directions from the body. Now, if the pseudopodia are stimulated with a rod, they begin to contract. The endosarc first passes inward, but the resistance of the body is so great that the fluid stops at the base of the pseudopodia. These, therefore, swell up in a bulbous fashion, as illustrated in Fig. 50. Such cases, indeed all the numerous cases in which the endosarc passes out of a pseudopodium and comes to rest as soon as it has left the latter, can only be explained on the assumption that the endosarc is forced out by the contraction of the ectosarc, or by some active movements of the endosarc itself, of a character not understood.

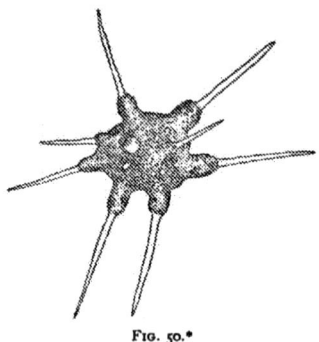

FIG. 50.*

Further, there are certain facts which speak positively in favor of the view that the production of the wrinkles is due to a contraction of the inner layer of the ectosarc. Thus, when an Amœba is strongly stimulated and withdraws all its pseudopodia quickly, the whole surface of the body becomes rough and wrinkled. The endosarc has not passed out of it, so that it cannot be considered in a state of collapse; on the contrary, it is clearly contracted as strongly as possible. Again, if a large pseudopodium is cut from the body, it contracts strongly, showing the rough, wrinkled contour, though the endosarc has not passed out of it.

---

* FIG. 50.—Specimen of *Amœba radiosa* in which the endosarc has passed out of the distal portions of the pseudopodia into the basal parts, causing them to swell up in a bulbous fashion.

The processes occurring in a retracting pseudopodium are the same as those taking place at the posterior end in a moving animal. In fact, the cases are really identical; the posterior portion of the body may be considered merely a large retracting pseudopodium. Often, as we shall see, it is impossible, from its method of formation, to consider it anything else.

The pseudopodia are, of course, usually formed in the anterior part of the Amœba, in contact with the substratum, and are directed in the line of progression, or at a slight angle with the line of progression. As they are withdrawn, their surface, as we have seen, usually becomes folded or roughened, and the small roughened projection resulting from the withdrawal lasts, as a rule, for a long time. As the Amœba continues its forward course, the base of the retracting pseudopodium retains nearly its original position, as do the other parts of that layer of the ectosarc that is against the substratum, so that the body of the Amœba gradually passes the pseudopodium, and the latter finally becomes united with the posterior end. During this transference to the rear, the pseudopodium usually changes its position (Fig. 51). At first it is directed nearly forward (*a*), then it takes a position at right angles to the body (*b*), and finally swings around with its point directed

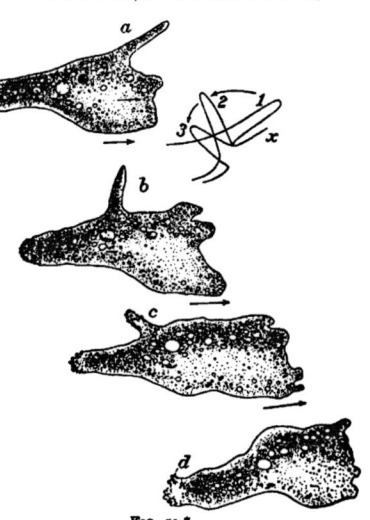

FIG. 51.

---

* FIG. 51.—Successive stages in the retraction of a pseudopodium. At *a* the pseudopodium extends forward at the anterior edge; at *b* it has partly withdrawn and stands at right angles to the body, which has partly passed it; at *c* the pseudopodium is directed backward, and is in the posterior part of the body; at *d* the small roughened remnant of the pseudopodium has nearly united with the tail. At *x* the successive positions occupied by the withdrawing pseudopodium are shown in a single diagram.

nearly backward (Fig. 51, c). This change of position is due to the contraction of the posterior part of the Amœba. The ectosarc just behind the base of the pseudopodium contracts toward the middle, as described on page 171. As a result the pseudopodium must swing around till it points nearly backward. The mechanism of the process will be best understood by an examination of Fig. 51, x.

Finally, what remains of the pseudopodium reaches the posterior end or tail (Fig. 51, d). By this time usually all that is left of it is a small roughened projection, its surface being of essentially the same character as that of the tail. This projection fuses completely with the tail, its projections taking up a portion of the surface of the latter. The tail is in fact nothing but the fused remnants of all the pseudopodia that have been formed, together with the contracted outer layer of the body of the Amœba (the latter cannot be distinguished in any essential way from a pseudopodium). A roughened tail is formed *de novo* whenever an Amœba suddenly changes its direction of movement. The previous anterior end then becomes roughened in contracting and forms a typical tail. This latter unites with the old tail if any of the latter remains. The substance of the tail gradually passes forward into the rest of the body, as we have seen.

MOVEMENTS AT THE ANTERIOR EDGE.

As in *Amœba verrucosa* and its relatives, so in the species of more changeable form, the most active movements are taking place at the anterior edge. In *Amœba proteus* and *A. limax* one sees still more distinctly than in the species before named the pushing forward of a series of waves of hyaloplasm which become attached to the substratum in front. In *Amœba limax* and its relatives especially such a wave may be very pronounced, extending forward at times one-fifth the length of the body or more, though usually much less.

At first the advancing wave, as it moves rapidly forward, is usually free from granules, and may be spoken of, therefore, as hyaloplasm. If the motion is less rapid, however, it contains granules, and is not distinguishable in any way from the interior endosarc. Where it is at first free from granules, it is nevertheless highly fluid in character, as is shown by the fact that it flows and spreads out swiftly, and that the granules of the endosarc pass into it rapidly. The freedom of the advancing hyaloplasm from granules is not due to its greater density or solidity as a result of the action of water upon it, as has sometimes been maintained, for it is at first free from granules; then, after the water has acted longer upon it, it becomes filled. Apparently the reason for its freedom from granules is merely the fact that it moves forward faster than the granules and leaves them behind. This view is supported by the variations to be observed in the relation of the clear substance to the granular substance at the anterior end. These variations arise chiefly from the different rates of movement of the two substances, and may be summarized as follows:

1. The clear substance moves fastest at first, and therefore becomes separated from the granules as a broad band in front (Fig. 52, *a*); this is then immediately filled completely by the granules. Even large granules or vacuoles pass to the very anterior edge, so that one sees but a line between these and the outer water.

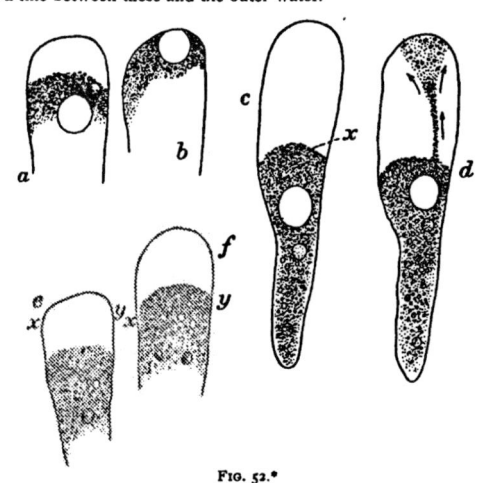

FIG. 52.*

2. The clear substance advances fastest, and so continues to do, so that it remains a long time as a broad, clear band in front of the granules.

3. The two substances advance at the same rate, so that there is no separation between them. The granules and vacuoles are then found at

* FIG. 52.—Distribution of hyaloplasm and granules at the anterior end in *Amœba limax* and its relatives: *a*, hyaloplasm without granules at the anterior end; *b*, granules and vacuole at the anterior edge; *c* and *d*, two successive instants in the locomotion; at *c* the anterior half of the body is free from granules, the latter being heaped up behind a well-marked barrier; at *d* the barrier has burst at a certain point (*x*), allowing a stream of granules to flow forward to the anterior edge; *e* and *f*, two successive stages at the advancing anterior end; in *e* the clear hyaloplasm has stopped at the line *x–y*; at *f* the hyaloplasm has advanced, while the granules are heaped up behind the same line *x–y*.

the advancing edge. This condition is not at all rare. In such cases there is at the anterior end less clear space between the granular region and the water than in any other part of the body. In fact, there is typically no space at all. I have seen large vacuoles come so close to the anterior edge at such times that it was not possible to distinguish between the boundary of the vacuole and the boundary of the Amœba (Fig. 52, *b*).

4. Finally, either hyaloplasm or endosarc or both may stop in any of the positions mentioned above. Thus the hyaloplasm may stop, whereupon the endosarc flows into it and fills it, or both may stop, so that the hyaloplasm remains empty, as a clear band, for a long time.

The line separating hyaloplasm and endosarc is at times very sharply defined, as has often been pointed out. A number of unusually favorable specimens gave me the opportunity of determining the reason for this, in many cases at least. The Amœba in question (Fig. 52, *c-f*) was an elongated, rapidly moving form, much resembling *A. limax*, but having usually two contractile vacuoles, one very large, in the fore part of the body, the other smaller and in the rear. The body contained many fine granules, which, when the animal was at rest, were scattered almost uniformly through the body; the peripheral, more solid zone (usually called ectosarc) contained as many of these as did the endosarc.

In moving this Amœba usually sends out first a large amount of clear fluid containing no granules; this at times extends so far as to constitute half the length of the Amœba (Fig. 52, *c*). There is a perfectly sharp line between the clear hyaloplasm and the granular endosarc, and behind this line the granules of the endosarc are heaped up, as if under pressure. Suddenly this line gives way over a small area (at *x*), and the granules pour through it in a thin stream nearly or quite to the anterior tip of the Amœba (Fig. 52, *d*).[*] Gradually the whole barrier gives way, and nothing is left to mark the position it occupied. If after its first outflow the hyaloplasm has stopped, the whole Amœba is filled with granules. But if, as is usually the case, after a pause the hyaloplasm has started forward again, the granules of the endosarc stream forward not to the anterior tip, *but only to the line which formed the anterior boundary of the hyaloplasm at its pause* (*x-y*, Fig. 52, *e, f*). Here the granules stop and are heaped up again, until finally the barrier breaks as before, and the granules rush forward again, to be stopped at a new line.

The explanation of these phenomena becomes evident on careful examination. It is to be noted that the line *x-y* which stops the flow of the granules of endosarc is always identical with one which

---
[*] Similar phenomena have been observed by Prowazek (1900, p. 17; Fig. 18).

formed the anterior boundary of the hyaloplasm (that is, of the Amœba as a whole) at a previous pause. The reason for this is as follows: The lower surface of the Amœba, as we know, is at rest; here the protoplasm has become modified to form a sort of membrane. This membrane extends up a very little distance at the sides and ends (as is shown by the fact that the protoplasm at the sides is at rest). Thus at the anterior end there is, after a pause, a low barrier formed by this membrane. The next wave of advancing hyaloplasm arises just behind this barrier, overleaps it, and pushes forward (the conditions being essentially the same as in *Amœba verrucosa*, already described). This advancing wave when first formed is very thin, forming a mere sheet lying on the substratum. This is shown by the fact that when the outline of the remainder of the Amœba is sharply in focus, the anterior portion is often undefined, and one is compelled to focus lower to get its outline sharply. The thin sheet of hyaloplasm which has just pushed forward is bounded behind by a low wall, formed from the membrane which previously limited it in front (Fig. 53, *x*). Now the granules of the endosarc flow forward until they reach this boundary; there they stop and become heaped up against it (Fig. 53). After a time the membrane forming this barrier, since it is now completely enveloped by protoplasm, becomes dissolved and gives way in the manner described above; the granules then flow forward. Meanwhile, a new partial boundary has been left in front by the hyaloplasm; this again stops the endosarc, and the whole process is repeated many times.

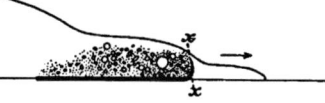

FIG. 53.*

Of course, when the anterior boundary advances uniformly, without pauses, no anterior membrane is formed, and there is nothing to hold back the granules of the endosarc; hence there is no reason for a separation of hyaloplasm and endosarc, and we find that none exists. On the other hand, when the Amœba has paused for a long time the anterior

* FIG. 53.—Diagram of a longitudinal section of the anterior edge of Amœba, to show the cause of the stopping of the granules of the endosarc some distance behind the anterior margin. The line beneath represents the substratum to which the Amœba is attached. The anterior hyaloplasm at first moves forward to the line *x-x*; stopping there it becomes covered with a firmer wall, as represented by the heavy black line. Now the hyaloplasm pushes forward from *above* the anterior edge *x-x*, forming a thin sheet closely applied to the substratum, as shown in the figure. The endosarcal granules flow forward, but are stopped by the barrier *x-x* (the former anterior boundary of the Amœba); they cannot flow forward till this boundary is liquefied.

membrane seems especially well developed, for the hyaloplasm pushes then a long way ahead and may form half the length of the Amœba before the endosarc has burst through the membrane.

The phenomena described above are very general in creeping Amœbæ, both in those with usually but a single pseudopodium (as *A. limax*), and in those with many pseudopodia (as *A. angulata*).

Of course, the general fact that there is a separation between hyaloplasm and endosarc is not explained by these observations; thus we know that they are often separated even in pseudopodia that are projected freely into the water. But the phenomena are much less marked in such cases; it is exactly the observed difference between them and the phenomena to be seen in a creeping Amœba that the above observations explain.

In all these details it is important not to lose sight of the essential point in the movement at the anterior end. This is as follows: The new wave begins on the upper surface just *behind* the former boundary line, and *rolls* forward, so that its former upper surface is now in contact with the substratum.

This method of movement explains a peculiar fact which one observes frequently, and which I found it difficult to understand before this movement was demonstrated. The advancing edge in Amœba usually does not push forward fine, loose particles lying on the substratum in front of it, but overlaps them instead, so that the Amœba creeps over them. This is especially noticeable when the water contains many particles of India ink or of soot. In view of the rolling movement with the series of waves, each coming from behind the previous anterior edge and thus descending on the substratum from above, this burying of loose movable particles becomes intelligible.

In *Amœba proteus* and its relatives the advancing anterior edge does not move forward in a single uniform curve, as it does in *A. verrucosa*, and, as a rule, in *A. limax*. On the contrary, the anterior edge may show the most varied form and modeling (Fig. 54); narrow points may be sent out here; a broad rounded lobe there; a rectangular projection elsewhere. Pseudopodia may rise above the general level, projecting freely into the water and later coming in contact with the substratum, or be withdrawn without coming thus in contact. The anterior portion of the body may divide into two lobes, or may become hollowed out so as to contain a cavity bounded by thin walls (see later, Figs. 75, 76). These facts show clearly that the method of advance of the anterior edge cannot possibly be determined by general pressure from behind, such as would be produced, for example, by a contraction of the posterior part of the body. Such pressure could not produce the cavity shown in Fig. 75 or the thin edge bearing numerous points shown in Fig. 54.

MOVEMENTS OF THE POSTERIOR PART OF THE BODY.

In an Amœba moving as a simple, elongated mass, the anterior portion of the body is broadest and very thin, being flattened against the substratum, while the posterior part is narrower and much thicker. In many cases the posterior end rises to an actual hump, the body being thickest at the posterior edge, or a little in front of this edge (Figs. 44, 58). This is true as well of the Amœbæ of nearly constant form (*A. verrucosa*, etc., Fig. 44), as of those related to *A. proteus*. From this hump the upper surface slopes forward to the thin anterior edge. The margins in the posterior part of the body are not thin, but rounded like the surface of a cylinder.

The anterior portion of the Amœba is attached to the substratum. This attachment of the anterior portion has been clearly demonstrated by Rhumbler (1898), and I can confirm his results throughout.† The attachment is probably by a mucus-like secretion; at least such a secretion exists, as Rhumbler and others have shown and I can confirm. I have sometimes been able to pull an Amœba about by using a glass rod to which a thread of this mucus had become attached (Fig. 55). The animal then seems to follow the rod at a distance, the thread of mucus not being visible. In virtue of this attachment the Amœba resists currents of water, or the impinging of solid bodies against it. The posterior portion of the body is not thus attached, but is entirely free from the bottom.

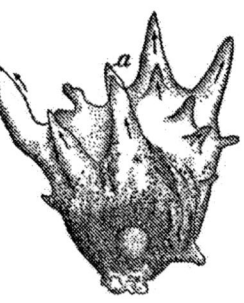

FIG. 54.*

In many cases the most posterior part of the body forms a more or less distinctly marked off portion, the surface of which is wrinkled or warty or villous, or otherwise irregular. This is variously known as the tail, the villous patch, or appendage (Wallich, Leidy), houppe (Penard), Schopf (Rhumbler), etc. The occurrence of this appendage is variable. In some species it is usually present, in others less common. Its occurrence and degree of development vary, indeed, in the same individual.

---

* FIG. 54.—*Amœba angulata* in locomotion, showing the numerous points in the anterior region, some attached to the substratum, others projecting freely into the water. *a* is the "antenna-like" pseudopodium, described on p. 177.

† It is rather curious that Bütschli (1892), in his discussion of the movements of Amœba, is inclined to deny that there is any attachment to the substratum.

In an advancing Amœba the posterior end moves forward at about the same rate as does the anterior, since the distance between the two remains about the same. Leaving out of account for the moment specimens with the wrinkled appendage, there is a continual current forward from the posterior end. Nevertheless, the latter remains on the average of about the same size. The material which flows out of it above is supplied from beneath. As we have seen, a layer of material at the under surface of the Amœba is at rest. The main portion of the body passes over this layer, dragging the posterior end. The latter takes up as it passes the resting layer which was against the substratum. This gradually becomes fluid, and passes forward again in the advancing current. All this may be clearly seen by observing the course of individual particles in the protoplasm and on the surface, and is fully set forth in the preceding pages of this paper.

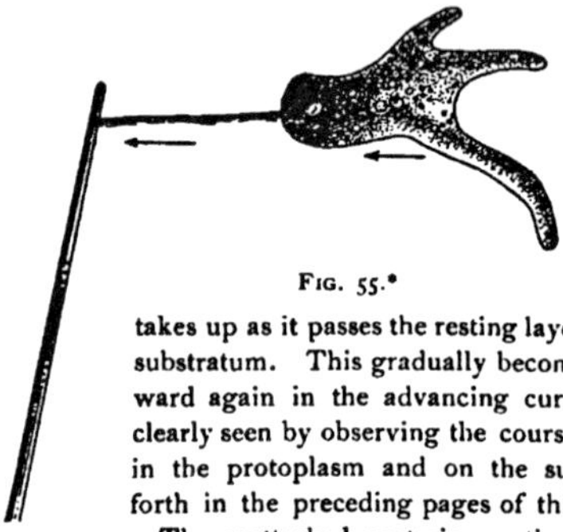

FIG. 55.*

The unattached posterior portion steadily contracts as it moves forward. Particles on its upper surface are moving forward, as we have seen in detail. But this is not all. Particles on its sides and under surface likewise move forward; there is an actual contraction independent of the streaming already described. The movement of substance due to this contraction is more striking and rapid as we approach the posterior end. As this contraction is an important fact, it will be well to give some details of the observations which show it to exist.

Particles attached to the lower surface, or to the lateral margins of the Amœba, next to the substratum, in the anterior part of the body, remain quiet for a long time. But this lasts only till they have reached that portion of the body which is free from the substratum; then they begin to move slowly forward as a result of the contraction just described. Of two such objects, one nearer the posterior end, the hinder one moves the more rapidly, so that the distance between the two slowly but distinctly decreases. Though such objects on the bottom or sides move forward, they do so less rapidly than does the posterior end. The latter, therefore, in time overtakes them, and they are finally pulled around the posterior end to the upper surface, where they pass forward, as we have already seen in detail.

* FIG. 55.—An Amœba drawn backward by a thread of its viscid secretion.

The contraction of the posterior part of the body is further shown by the behavior of retracted pseudopodia. When a pseudopodium contracts it usually produces, as we have seen, a small wart-like excrescence, which persists for some time. Such wart-like remains of pseudopodia behave like foreign bodies attached to the margin of the Amœba. In the anterior half they remain quiet, while in the free posterior half they move slowly forward, as a result of the contraction of this part of the body. When two or more of these remnants of pseudopodia are formed at once, with an interval between them, this interval becomes less as a result of the more rapid movement of the hinder one.

The contraction of this posterior region is sometimes very striking, especially when the posterior end becomes attached to some foreign object and is drawn out longer than usual; when it finally becomes free it contracts suddenly and rapidly. Thus, for example, an Amœba having the form shown in Fig. 56, *a*, began to move in the direction shown by the arrow, when it became evident that the posterior end was attached to a diatom shell, which was fast to the substratum. As the Amœba crept away the posterior end was drawn out, as shown at *b*. Finally the diatom became detached from the bottom, when the stretched posterior end at once contracted, shortening up rapidly, so that the Amœba had the form shown at *c*. Such observations are often made.

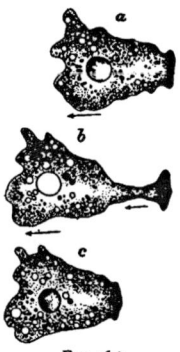

This contraction does not occur in that part of the Amœba which is attached to the bottom, but begins at once as soon as the attachment ceases. One might compare the outer layer to a stretched sheet of India rubber that is attached to a surface by means of some adhesive substance. As soon as the adhesion gives way the sheet contracts. There is no definite point at which the attachment to the substratum must cease; sometimes it is farther forward, sometimes farther back. The gradual freeing of the posterior portion can be clearly observed in many cases, particularly in *Amœba angulata*, and may be seen to go hand in hand with the contraction of the ectosarc. As soon as a certain part of the body becomes free, its contraction takes place with some suddenness, and the contraction is the more noticeable the greater the part of the body that is freed at once. Often the process of becoming freed

---

* Fig. 56.—Successive forms of an Amœba, showing the marked contraction of the posterior end. (See text).

takes place in a series of jerks, and there is a corresponding jerkiness in the contractions of the posterior part of the body. When a large amount of surface is freed at once there is a sudden forward rush of the fluid portion of the Amœba, with a striking contraction of the posterior part of the body. Such a case as is shown in Fig. 56 is only an unusually strong contraction of this sort, due to the fact that the hinder part on the body had remained locally attached longer than usual.

As a result of this contraction, the ectosarc of the posterior part of the body becomes thickened and wrinkled, or warty. The change from a flat plate to a rounded form involves a decrease in the amount of external surface, and as the amount of material in the surface layer is not at once decreased, this layer is compelled to fold and become wrinkled and warty. When this process is very pronounced we have produced at the posterior end the wrinkled, warty appendage so often described. Such a roughened structure may be produced in any part of the Amœba by rapid contraction, as we have seen above (p. 160). The rough, warty appendage at the posterior end is the common product of all the contractions which have taken place.

In *Amœba verrucosa* and its relatives the current forward on the upper surface extends backward to the posterior end; the outer surface of the latter seems not markedly different in texture from that of the rest of the body. In other species of Amœba there is a greater difference between the texture of the surface layer of the anterior part of the body and that of the posterior end, and this may involve some differences in the movements. Often, in even these species, the forward current extends backward to the very posterior end; particles on the under side pass up over the posterior end and forward, just as in *A. verrucosa* (see p. 147). But in other cases, in *A. limax*, *A. proteus*, etc., the surface material at the posterior end is so stiffened that it is temporarily excluded from the current. There is then produced the distinct, roughened appendage, which is for a time dragged passively behind the Amœba. In such a case the currents from beneath pass upward on either side of this appendage, meeting in the middle line (Fig. 57). Particles attached to the under surface on either side of the appendage, therefore, soon pass to the upper surface and are carried forward, while those on the under surface of the appendage itself may remain in position and be dragged forward for a considerable time.

But I have rarely found this posterior appendage so completely cut off from the general circulation as is often supposed. Usually there is a very slow current forward on the upper surface of the appendage, involving also its internal parts. Into this current particles attached to the posterior end, and even to the under surface of the appendage, are in course of time drawn. I have thus seen particles of soot dragged about for ten minutes at the posterior end, then finally pass upward into the surface current, where they were carried to the anterior end. Even when this slow current from the posterior appendage is completely suspended the suspension is only temporary. The currents after a period begin again, and a strongly marked warty posterior appendage may in time completely disappear, its substance having become mingled with that of the remainder of the Amœba.

Other parts of the Amœba may become temporarily immobile and thus excluded from the general circulation. One often sees certain parts of *Amœba proteus* and other similar species thus quiet, while the rest of the body is in active motion (see Rhumbler, 1898, p. 122).

FIG. 57.*

In the species with "eruptive" pseudopodia this process seems to have gone farthest. Here the whole outer layer apparently becomes hardened at times, so that movement occurs only when the inner substance bursts through this, forming "eruptive" pseudopodia. In this case the pseudopodia formation apparently differs from that in *Amœba proteus* and its relatives, as described above, in the fact that the ectosarc of the body is not transferred to the surface of the pseudopodium. I have not been able to study this process in detail further than to determine that there is no backward current on such pseudopodia. The matter is worthy of further examination. In *Amœba verrucosa*, the species which has been hitherto supposed to have the most immobile ectosarc, we have shown that the outer layer is in continual rotary motion in the progressing animal.

GENERAL VIEW OF THE MOVEMENTS OF AMŒBA IN LOCOMOTION.

Let us now, with the aid of a diagram, attempt to form a clear conception of the movements occurring in an Amœba that is progressing in a definite direction, with a view to determining the nature of the forces at work. Fig. 58 may represent a longitudinal section of such an Amœba as seen in a side view. The anterior end, A, is, as we have seen, very thin, and is applied closely to the substratum, while the posterior end, P, is high and rounded, forming a sort of pouch. It is free from the substratum, beginning at the point x, at about the middle of the body.

At the anterior end waves of hyaloplasm are pushed forward one after the other, so that the anterior end successively occupies the positions a, b, c. As we know, it is the upper surface which thus pushes out; it rolls over, so that a point which was originally on the upper

* FIG. 57.—Diagram of the surface currents when the posterior appendage is excluded from the general stream.

surface becomes applied to the substratum. It is evident that the surface of the Amœba is increased in extent by the pushing out of these waves.

On the upper surface of the Amœba there is a forward current of the outer layer, as indicated by the arrows. This current extends backward to the posterior end, where it is continued as a movement upward from the lower surface. This upward movement stops, at any given movement, at about the point y, though, of course, the point where it ceases cannot be precisely fixed. That part of the lower surface which is in contact with the substratum, from the anterior end to x, is quiet. That part of the lower surface which is not in contact with the substratum (from x to y) is moving slightly forward owing to the contraction in this region, as described on pages 166–168. This movement is comparatively slight, as indicated by the small arrows. Within the

FIG. 58.*

Amœba are currents moving forward in the same direction as the current on the upper surface.

The posterior end as a whole moves forward, so that it comes to occupy later the position shown by the dotted line. The point x, where the lower surface of the Amœba becomes free from the substratum, moves forward an equivalent amount to x'. The entire Amœba thus moves forward in the direction indicated by the large arrow above.

Thus far our account has been purely descriptive, containing only what has been demonstrated by observation and experiment, and introducing nothing hypothetical. We must now endeavor to form a conception of the location and direction of action of the forces at work in producing the movements. Discussion of the ultimate character of the forces will be reserved for the section on the theories of amœboid movement.

One of the primary phenomena is evidently the pushing forward of

* FIG. 58.—Diagram of the movements of Amœba, as seen in side view. A, anterior end; P, posterior end; a, b, c, successive positions occupied by the anterior edge. The large arrow above shows the direction of locomotion; the other arrows show the direction of the protoplasmic currents, the longer arrows indicating the more rapid currents. For further explanation see text.

the anterior edge. The form of the Amœba shows that this cannot be due to pressure from behind, for if the pressure were greatest behind and less in front, the mass of internal fluid would, of course, be forced forward, and the Amœba would be thickest in front instead of behind; the form of anterior and posterior ends would be reversed. In the varied modeling of the anterior end we have seen another proof of the impossibility of accounting for the action here by pressure from behind (p. 164). Further, the forward current on the upper surface of the Amœba could not possibly be produced by pressure from behind. The impossibility of accounting for the form and movement by pressure from behind has been recognized by most investigators, though on other grounds than those here set forth.

On the other hand, if we take the view that the anterior wave, after attaching itself to the substratum, exercises a pull on the parts behind it, the rest of the phenomena follow most naturally. Such a pull would draw forward the tenacious upper layer of the body, and we find that this is actually moving forward. The mass of inactive internal contents would drag behind as far as possible, so that the thickest part of the Amœba would be at the rear, and this is exactly what we find to be true. The posterior end would be dragged forward. This, also, is true. In its movement forward it would be partly rolled; that is, its lower surface would gradually pass upward and become the upper surface. This, also, we know to be true.* Finally, the internal fluid contents would be compelled to stream forward as the anterior end advanced and the posterior end followed it. This streaming is, of course, one of the striking features of the Amœba. The characteristics of the endosarcal streaming are, I believe, exactly what might be expected from the method of origin just set forth.

We have one other more or less independent factor of the movement in the contraction of the posterior part of the body that is not in contact with the substratum. As we have seen, the substance of the sides and bottom, as well as of the upper surface, are moving forward in this region, as indicated by the small arrows of the diagram (between x and y). Moreover, we know that there is a lateral contraction as well as an antero-posterior one, for the wide, flat anterior portion shrinks together as soon as it is released from the substratum. This contraction should probably be brought into relation with the previous increase of surface at the anterior end. As the anterior wave is sent forth the surface at the anterior end is much increased. We might compare the action with the stretching of a sheet of India rubber. This tense portion then becomes attached to the substratum, as we might

---

* Large bodies within the Amœba close against the posterior surface are often rolled over in this process, as I have several times observed.

fasten the stretched sheet of India rubber by means of a strong cement to a plate of glass. Upon being released the tense layer of protoplasm contracts again, just as the rubber sheet would do. In the protoplasm the contraction takes place rather slowly, causing the steady pulling forward of the posterior half of the body, as exhibited by objects attached to its lower surface.

In connection with the contraction of the lower surface as just described we must consider also certain properties of the upper surface. When this is pulled upon by the advancing anterior wave it does not respond like an inelastic membrane, but like an elastic, contractile one. If it were inelastic its motion would follow that of the anterior end exactly, and thus take place in a series of jerks. But this does not occur. When the anterior wave pushes forward, thus extending the upper surface, there is no immediate increase in the movement of this surface, nor of the posterior end; the movement forward is a steady one. The property of contractility is further shown very directly in the phenomena which take place when a large portion of the lower surface of the Amœba is suddenly released, as described on page 167. It seems clear that the entire surface of the Amœba is in a state of tension and that this tension is directed toward the advancing anterior end. The condition would be imitated by partly filling a rubber sack with a heavy fluid, causing one surface to adhere to the substratum and pulling on one side.

As a result of this tension on the surface the internal contents of the Amœba must, of course, be under a certain slight amount of pressure. As we have seen (p. 171) even without this pressure the internal contents must flow forward, since the posterior surface, against which they are resting, is moved forward. But the pressure accounts for certain details of the movements of the endosarc. Thus, when a pseudopodium is sent forth, or one of the anterior waves moves forward, it is usually soon filled by endosarc. In its pushing forward the pseudopodium forms a region where the tension is relieved; the fluid contents, under pressure elsewhere, therefore flow into it.

It is to be noted that this pressure is a mere consequence of the tension due to the pushing forward of the anterior edge, and is by no means a cause of the pushing forward; it is always, therefore, subordinate to and dependent upon the latter, and is not a matter of primary significance.

Altogether, then, our results lead us to look upon Amœba as an elastic and contractile sac, containing fluid. In locomotion one side of this sac actively stretches out, becomes attached to the substratum, and draws the remainder of the sac after it in a rolling movement. The primary phenomena are the stretching out of one side, the elasticity, and the contractility of the outer layer.

Whether this elasticity and contractility should not be considered properties, not merely of the outer layer, but of the entire substance of the Amœba, may be a question. The fact that ectosarc and endosarc are mutually interconvertible would seem to imply an affirmative answer to this question, and I believe other evidence could be adduced looking in the same direction. But the locomotion itself seems to require these properties only in the ectosarc, so that we shall for the present leave out of consideration the question as to their existence in the endosarc.

To the three primary phenomena above mentioned we must devote further attention. It has been maintained by certain writers that the ectosarc is not an elastic and contractile membrane, as above set forth; hence we must examine the evidence on that point. There then remain the questions: What is the cause of the pushing out at the anterior edge? and, What is the essential nature of the contractility of the ectosarc? These questions will be reserved for a special section on the theories of amœboid movement. We will at this point investigate certain general properties of the substance of Amœba, with a view especially to determining whether we are justified in considering the ectosarc elastic and contractile, though not limiting our attention to these properties alone.

SOME CHARACTERISTICS OF THE SUBSTANCE OF AMŒBA.

FLUIDITY.

It is, of course, not necessary to dwell upon this point; it has been treated in detail recently by Rhumbler (1898, 1902) and Jensen (1900). For anyone who is familiar with the movements of Amœba from personal observation, doubt cannot exist that its protoplasm has at least some of the most striking properties of fluids, notably the property of flowing, with the freedom of movement of the particles with reference to each other that this implies. This applies most strongly to the endosarc; for the ectosarc, as we shall see, there are decided limitations to the fluidity. Nevertheless, the particles of the ectosarc have, to a considerable degree, that freedom of movement with relation to each other that is characteristic of fluids. This is shown, for example, by the fact that any portion of the ectosarc may be temporarily excluded from the advancing stream (especially common at the posterior end, p. 169), and in the fact that neighboring portions of the ectosarc may flow in opposite directions (p. 148). But the characteristics of the ectosarc are much more those of a tough, rather persistent skin than has sometimes been supposed. This point is brought out in the following sections.

RHUMBLER'S "ENTO-ECTOPLASM PROCESS."

The movements of the outer layer of the body described in this paper have an important bearing on the transformation of endosarc into ectosarc and *vice versa*, of which so much is said in Rhumbler's recent extensive paper (1898). My observations show that this transformation is confined within much narrower limits than Rhumbler supposed. In the account which he gives of the movements of Amœba this transformation (the "ento-ectoplasm process," as he calls it) plays a very large part, and is essential to locomotion. At the anterior end, according to Rhumbler, endosarc constantly flows out to the surface and is there transformed into ectosarc, flowing back as such along the surface of the body. Somewhere in the posterior part of the body or at the base of a pseudopodium this ectosarc passes inward and is retransformed into endosarc. These supposed processes are indicated in the diagrams from Rhumbler, Figs. 30-33.

My observations show that this view of the constant inter-transformation of the two layers is incorrect, and that we must attribute to the ectosarc a much higher degree of permanence than Rhumbler supposed. There is no regular transformation of endosarc into ectosarc at the anterior end. On the contrary, the ectosarc here retains its continuity unbroken, moving across the anterior end in the same manner as across other parts of the body. In the same way, the ectosarc is not regularly transformed into endosarc in the hinder part of the body. We can trace a single definite point on the ectosarc (or a complex of such points having a definite relation to each other) continually until it has passed completely around the Amœba; several complete rotations of this sort are described from actual observation on page 141. In the species having very changeable forms a single point on the ectosarc may be traced, for example, from the surface of a pseudopodium at the posterior end across the whole length of the body to near the tip of a long pseudopodium at the anterior end (Fig. 49, p. 155); there is no reason to suppose that it could not be traced indefinitely but for the difficulties of observation.

On the other hand, there is no doubt that ectosarc may be transformed into endosarc, and *vice versa*, under certain conditions. This process was apparently first clearly seen by Wallich (1863, *a*, p. 370). Wallich saw the formation of "eruptive pseudopodia" by the outflow of the endosarc through a small perforation in the ectosarc. A portion of the latter was thus covered by the endosarc, and gradually resorbed. Rhumbler figures a similar case (1898, p. 152), and I have repeatedly seen such. Further, as Rhumbler has shown, and as I can confirm, in *A. verrucosa* food bodies are enclosed in a layer of thick ectosarc, which passes with the food into the endosarc, there to be resorbed (see p. 195).

Thus, it is clear that there may be a transformation of one layer into the other under special circumstances, but such transformation is much less general than Rhumbler supposed, and is by no means a regular accompaniment of locomotion.

ELASTICITY OF FORM IN AMŒBA.

For a real understanding of the phenomena shown by amœboid protoplasm it is important to realize that it has, besides some of the chief characteristics of fluids, a number of properties that are usually considered characteristic of solids. This came out clearly in certain of my experiments. They show that Amœba has elasticity of form to a considerable degree.

These experiments consisted in changing the shapes of Amœbæ with a fine capillary glass rod, under the microscope, in the open drop. From numerous experiments of this character the following may be selected as typical:

An Amœba had sent out one rather long, thick pseudopodium, as shown in Fig. 59, *a*. With the capillary glass rod this pseudopodium, *a*, was pulled loose from the bottom and bent over into the position shown by the dotted outline *b*. On being released it rather quickly

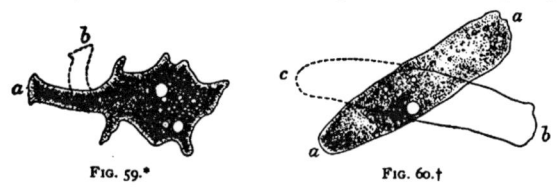

FIG. 59.*   FIG. 60.†

sprang back into its original position, *a*. This experiment was repeated on different Amœbæ many times.

An elongated Amœba (*a–a*, Fig. 60) was bent with the rod at about its middle, so that the anterior half was pushed far to one side of the original median axis (to *b*). This anterior half at once attached itself to the bottom, whereupon the posterior half, which was not attached, immediately swung round into line with it, so that the Amœba occupied the position *b–c*. Thus the original straight Amœba on being bent immediately straightens itself out again. On repeating this experiment with many elongated individuals it was found that frequently the straightening out was not quite complete, so that after it had occurred there was still a slight bend in the middle.

An Amœba had a long pseudopodium curved over to one side, as in Fig. 61, *a*. This pseudopodium was loosened from the bottom with

---

*FIG. 59.—A straight pseudopodium, *a*, is bent into the position *b* with a rod. It at once returns to the position *a*.

†FIG. 60.—A narrow Amœba, *a–a*, is bent with the rod into the position *a–b*, the end, *b*, then becomes attached, and the animal at once straightens into the position *b–c*.

the rod and straightened out (*b*). On being released it at once swung back to its original form and position at *a*.

An Amœba had many long, slender, free pseudopodia standing out radially from the body. These could be pushed repeatedly to one side or the other, or bent at a marked angle. In every case they returned at once to the radial position.

An indefinite number of such experiments could be detailed. They show clearly that Amœba has, to a certain degree at least, one of the most distinctive properties of solids, a tendency to resist deformation of shape, and to restore the shape when changed. It will be observed that the cases are not such as can be accounted for on the assumption that Amœba is a simple fluid mass which tends to take a certain form in accordance with the principle of least surfaces. A small sphere of fluid when deformed returns to its original shape in conformity with the principle just mentioned. But such returns to the original form as are illustrated in Fig. 59 and Fig. 61, for example, are not required by the principle of least surfaces so long as the Amœba is considered a simple fluid mass.

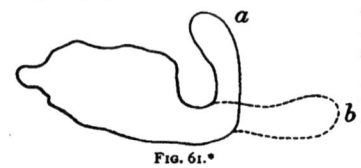

FIG. 61.*

On the other hand, if we consider Amœba not as a simple fluid, but as a fluid mass of a foamlike or alveolar structure, composed of a tense meshwork of one fluid enclosing minute drops of another, then the results above set forth might be explained without assuming that the protoplasm has in any part passed from a liquid to a solid state. This follows from the considerations and experiments recently set forth by Rhumbler in a most important and suggestive paper (1902). Rhumbler shows that a fluid mass having alveolar structure must react to transient pressure from without like an elastic body; in other words, that it must have elasticity of form. The results which I have set forth above might almost seem, then, to have been predicted in Rhumbler's summing up: "Transient tensions or pressures produce an elastic reaction of the cell body; longer action, on the other hand, produces a plastic reaction" (*l. c.*, p. 371). For the detailed demonstration of this principle the reader must be referred to the original paper of Rhumbler. It will suffice to note here that the result is due to the fact that deformations of the body as a whole produce deformations of the alveoli, and that the surface tension of the alveolar walls tends to

* FIG. 61.—A large, curved pseudopodium, *a*, is straightened out into the position *b* with a rod. On being released it at once returns to the position *a*.

restore them at once to their original form, resulting in a return of the whole body to its original form.

If, then, we hold that the substance of Amœba is composed of such an alveolar structure, the above observations are intelligible, even though no part of the substance of the organism is in the solid state. But whatever the cause, we must recognize that the protoplasm of Amœba shows in the gross some of the characteristics of solids.

Leaving out of account the minute structure of the protoplasm, most or all of the observations detailed above could be accounted for on the assumption that the body of Amœba consists of a sac of fluid, the outer wall of the sac being tough and elastic, and, as is shown later, contractile. In this case only the outer layer, the ectosarc, would show the characteristics of matter in the solid state of aggregation. Certain observations made by the writer indicate that this is the true state of the case. These observations are as follows: In several cases a long, slender pseudopodium, formed of both endosarc and ectosarc, was stimulated at the tip, causing the endosarc to be withdrawn, and leaving the pseudopodium formed of ectosarc alone, as illustrated in Fig. 50, page 158. Such pseudopodia could with the glass rod be bent sharply at an angle, and would often remain thus for some time. If, while thus sharply bent, the endosarc, as sometimes happens, begins to flow back into the pseudopodium, the latter straightens out with a sort of jerk as soon as the endosarc begins to fill it. The ectosarc thus acts like an empty glove-finger, which might bend over when empty but which would straighten out on becoming filled with a fluid. A tough skin could perhaps be formed by an alveolar fluid in accordance with the principles developed by Rhumbler as above set forth. Whatever the explanation, the experiments indicate the existence of this tough skin-like layer on the outer surface of the body.

CONTRACTILITY IN THE ECTOSARC OF AMŒBA.

Besides elasticity of form, the outer layer of Amœba clearly has the power of contracting locally. This is a fact which is omitted from consideration in many of the theories in which amœboid movement is referred to local changes in the surface tension of a fluid mass. It will be well, therefore, to set forth some of the observations on this point. I transcribe here some of my notes, with the corresponding sketches.

1. Specimen with a single long, prominent, curved pseudopodium. This rather quickly swings around bodily toward its concave side, and unites with the protoplasm of the body (Fig. 62, *a*).

2. A specimen sends out a long, curved pseudopodium (Fig. 62, *b*). This slowly straightens out, passing from position 1 to position 2.

3. *A. angulata* usually sends out at the anterior end a single pointed pseudopodium obliquely upward into the water (Fig. 62, *c*). This point frequently waves from side to side like an antenna.

4. An Amœba was creeping on the surface film, with a very long, slender pseudopodium trailing behind down into the water and bent to one side. This pseudopodium suddenly swung far over to the other side.

5. Amœba with many pseudopodia extending in all directions freely into the water. Just as withdrawal begins, a given pseudopodium bends over to one side, becomes curved to form a half circle, or waves back and forth from one side to the other.

An indefinite number of such observations could be adduced, showing that with its other movements Amœba has the power of bending and straightening its pseudopodia and waving them from side to side. Such movements have, of course, been described by many authors; in the magnificent monograph of the Rhizopoda by Penard (1902) many still more striking cases than those I have described are set forth. Some of them should be quoted. In *Amœba radiosa* the pseudopodia "may be displaced as a whole in the liquid, and I have seen them describe in this manner a quarter of a full circle in a second, like the hands of a watch which one pushes forward suddenly by fifteen minutes. On two or three occasions also I have noticed in the very sharp point of a pseudopodium a rapid movement of wave-like vibration, so that one could compare it with a flagellum" (*l.c.*, p. 88). Similar phenomena are described for *Amœba limax* (p. 36), *A. gorgonia* (p. 79), and especially for *A. ambulacralis* (pp. 91, 92), in which the pseudopodia act like tentacles. In other rhizopods, relatives of Amœba, Penard describes similar phenomena. Thus in *Pamphagus mutabilis* (p. 439) the pseudopodia are said to move as a whole in the water "almost as quickly as flagella." Similar facts are described for *Difflugia pristis* (p. 255), *Cystodifflugia sacculus* (p. 429), *Pamphagus granulatus* (p. 436), *Nadinella tenella* (p. 462), and various other rhizopods. Penard compares the movements

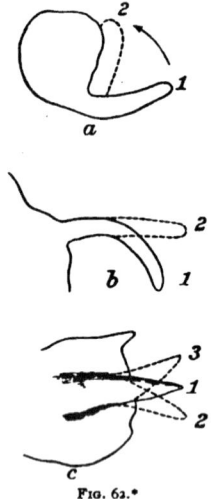

FIG. 62.

---

* FIG. 62.—Movements of pseudopodia: *a*, a pseudopodium in the position 1 bends quickly in the direction shown by the arrow, and unites with the body; *b*, a curved pseudopodium, 1, straightens into the position 2; *c*, the antenna-like anterior pseudopodium of *Amœba angulata;* it vibrates from 1 to 2, thence back through 1 to 3, etc.

of the pseudopodia in many cases to the vibrations of flagella. Similar movements of pseudopodia have, of course, been described by other authors, including Bütschli (1878, p. 272; 1880, p. 123). The striking resemblance of the movements of the pseudopodia in some cases to those of flagella (see especially the account of Podostoma, Claparède & Lachmann, 1858, p. 441) seems to indicate that the motion in these two classes of structures must be essentially similar in character, and that no theory of amœboid movement is likely to be correct that is inconsistent with the movements of flagella. Certain suggestions as to the possibility of bringing the two in relation are given in the theoretical portion of the present paper (p. 218).

The whole body is sometimes moved rapidly by such movements of the pseudopodia. This happens especially when the body is suspended in the water and bears many long pseudopodia, one of which comes in contact with the substratum. This pseudopodium spreads out and extends along the surface for a distance, the part along the surface forming nearly a right angle with the free portion. Suddenly the pseudopodium straightens; since the distal end is attached, the body is thrown almost violently against the substratum.

Somewhat similar movements take place frequently in *Amœba verrucosa* and its relatives, without the formation of pseudopodia. The course of events is usually as follows: A specimen is creeping in a certain direction in the usual manner with the anterior border attached, while the posterior end is raised a slight distance from the substratum. As a reaction to stimulus, or for some other reason, the anterior end releases itself from the bottom. The posterior end thereupon sinks down and becomes attached. Then its ectosarc contracts slightly, in such a way as to lift the anterior end suddenly. The animal thus stands upon what was its posterior end. Now, by varied contractions of the parts of the ectosarc in contact with the substratum, the animal may jerk from side to side rapidly and repeatedly, reminding one of the movements of certain caterpillars which jerk their anterior ends about in a similar manner. These movements are very striking and are much more rapid than any that occur in other species of Amœba, so far as I have observed.

The animal may even move from place to place in this manner. Standing on one end, it jerks its body suddenly over to one side, so that the previously upper end comes close to the substratum. This end now becomes attached, while the other is released. Next a new sudden contraction brings the released end upward, so that the animal now occupies a new location, one body's length from that previously occupied. I have never seen the movement go any farther than what I have just described, so that there is no evidence that this method is employed for bringing about orderly locomotion.

These movements remind one of the "rolling motion" described by Rhumbler (1898, p. 115) for these species, though they take place without any noticeable change of form and in a manner entirely different from the movements described by Rhumbler. As we have seen above (p. 140), the normal locomotion of these species is, in a certain sense, of a "rolling" character, so that the phenomena described by Rhumbler as the rolling movements, perhaps really presented nothing different in principle from the usual motion, though occurring in a different way because the organism was unattached.

In addition to movements of the character above described, certain other phenomena show in a different way the contractility of the ectosarc. Thus, I stimulated sharply with a glass rod one side of an elongated moving specimen of *Amœba limax* about one-third its length from the posterior end (Fig. 63, *a*). The body at once contracted rapidly, in a ring-like manner (*b*), at this point, and in about 1½ seconds the posterior portion was cut off completely, save by a fine thread (*c*), by which it hung to the anterior portion for a minute or two. Later this broke, and the posterior piece finally underwent degeneration.

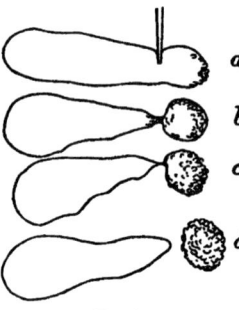

Fig. 63.*

Penard, in his great work on the Rhizopoda, describes similar phenomena in *Amœba terricola* (*verrucosa*) after injury to the ectosarc. After a small injury the injured region is invaginated, forming a small tube passing inward, which is later resorbed. But if the injury is large the part surrounding it contracts strongly, forming a deep constriction between it and the remainder of the body (Fig. 64), and this injured portion is finally constricted off completely (Penard, 1902, p. 109).†

Altogether, then, we may consider it thoroughly demonstrated that the ectosarc has the power of contracting in definitely limited regions in such a way as (1) to produce movements of entire pseudopodia comparable to those of flagella; (2) to produce ringlike contractions which may even progress so far as to cut the body in two completely.

We need not, therefore, hesitate to admit the existence of contractions of the ectosarc in ordinary locomotion; these are, for the rest, as clearly observed as those just described.

* Fig. 63.—An *A. limax* is stimulated strongly near the posterior end at *a*; the stimulated part thereupon constricts (*b*, *c*), separating off the posterior end (*d*).
† For other observations on reactions to injuries see pp. 202–204.

REACTIONS TO STIMULI.

Of particular importance for the understanding of the behavior of organisms are those reactions which determine the direction of locomotion. Experiments show that the stimuli to such reactions must, in a slow-moving organism like Amœba, affect only one side of the body, or at least affect different parts of the body differently. Owing to the minute size of Amœba, it is difficult to apply stimuli in such a way as to fulfill this condition. Heat or cold, or a chemical in solution, for examples, when applied to one side are likely to extend to the other side, and far beyond, before the slow reaction of Amœba has taken place; the reaction when it occurs is then to a general, and not to a local stimulation. For this reason the reactions of Amœba to such general stimulation are much better known than those to stimuli locally applied. I have devoted myself to the reactions to localized stimuli, and have succeeded in overcoming the experimental difficulties for a number of different classes of agents, though not for all.

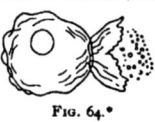

FIG. 64.*

In examining the reactions to stimuli, it will be necessary to keep in mind the method of locomotion (set forth briefly on p. 169; diagram in Fig. 58, p. 170). The factors to which special attention must be paid are: (1) the sending out (or rolling over, as perhaps it would be better to say) of waves of the ectosarc on one side, determining the anterior end in the locomotion; (2) the attachment to the substratum; (3) the contraction of parts of the body.

The reactions were studied chiefly in *Amœba proteus* and *A. angulata;* where other species were used, they are specifically mentioned.

REACTIONS TO MECHANICAL STIMULI.

The reaction to mechanical stimuli may be either positive or negative.

POSITIVE REACTION.

An Amœba floating in the water frequently takes a starlike form, with many long pseudopodia projecting in all directions. If one of these pseudopodia comes in contact with a solid object or the surface film (which may always be considered a solid for these purposes), the portion in contact flattens out, attaches itself to the object, and its protoplasm begins to flow out in a sheet over the latter. The other pseudopodia are now slowly withdrawn and the entire animal spreads out on the solid, moving usually in the direction inaugurated by the first pseudopodium which came in contact. Often in passing to the surface of the solid there are a number of rapid jerking movements, due to

---

* FIG. 64.—*A. verrucosa* constricting off an injured region, after Penard (1902).

straightening out or bending of pseudopodia as described on p. 179. After becoming completely transferred to the surface of the solid the form may differ much from that of the floating Amœba. Fig. 65 illustrates such a reaction. A floating Amœba will thus spread out on the substratum, on the surface film, or, so far as possible, on small masses of débris suspended in the water.

An Amœba which is moving along a surface also shows at times a positive reaction to mechanical stimuli by turning toward small objects with which it comes in contact at one side of the anterior end. This reaction takes place very frequently in the normal locomotion of Amœba, but I have not been able to produce it experimentally by touching one side of the animal with a glass rod. This is because it is difficult to give a touch so light that it shall not induce the negative reaction. I shall give a detailed account of reactions that probably belong here in connection with the account of food reactions (pp. 196–202, and Figs.

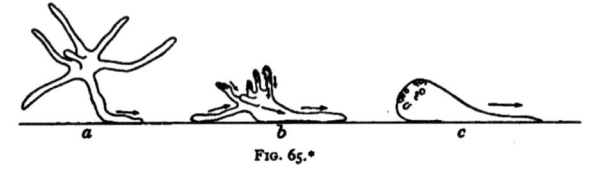

FIG. 65.*

73–76). As will there be shown, the reaction is often long continued and rather complicated.

Le Dantec (1895) gave a good account of the positive reaction of Amœba, as shown in its spreading out on solids.

NEGATIVE REACTION.

I have studied the negative reaction to mechanical stimuli by touching a spot on one side or end of the animal with the tip of a fine glass rod. A glass rod may easily be so drawn out that its tip is as fine as the tip of a pseudopodium, and with some practice it is possible to give, under the microscope, in the open drop, very precisely localized stimuli with this.

We will first examine the reaction to a rather strong stimulus at the anterior edge of an Amœba that is creeping forward with outspread anterior end and contracted posterior end in the usual way. The tip of the glass rod is thrust sharply against the anterior edge, producing

* FIG. 65.—Positive reaction to a mechanical stimulus in Amœba, in side view. A floating Amœba comes in contact by one of its pseudopodia with a solid (*a*); it thereupon passes to the solid, withdrawing the other pseudopodia (*b* and *c*). See text.

a depression in the ectosarc that may last for some time (Fig. 66). (We will suppose that the thrust does not detach the Amœba from the surface, as sometimes happens.) At once the anterior portion of the Amœba ceases to advance. It remains quiet for a definite interval, which I should judge to be about a second, while the current from behind continues to move forward. As a result of the stoppage at the anterior edge there is a heaping up of the protoplasm in the middle of the body. After about a second the part stimulated begins to contract and currents start backward from it. Thus the currents from the two ends meet in the middle, often producing a further heaping up in this region. Usually, however, the ectosarc of one side of the Amœba quickly gives way and a new pseudopodium starts out laterally. As a rule this new pseudopodium is formed near the original anterior margin, often at the very edge of the area directly affected by the stimulus (Fig. 66). The reason for this is evident. Only this anterior half of the Amœba is expanded and attached to the substratum, the posterior half being free and contracted. It is, therefore, much easier to continue locomotion by sending out pseudopodia somewhere in the attached region than behind it. If sent out in the unattached region, the original contraction would have to be overcome, and no locomotion could occur until the pseudopodium had (by chance?) come in contact with the substratum and become attached to it. By sending out pseudopodia thus in some portion of the attached region, the movement is, in a certain sense, a continuation of that which was taking place before stimulation, though in a different direction. The Amœba follows a path which forms an angle with its previous one.

The course of the reaction may vary considerably from that above described. If the stimulus is weak the reaction may consist merely in a stoppage at the point stimulated without any contraction there. The current from behind continues; a pseudopodium breaks out at one side of the region stimulated, and the Amœba moves in the direction so indicated. If the stimulus is *very* weak the current may cease only for an instant in the region stimulated, then continue as before; the direction of progress thus remains unchanged.

If the stimulus is very strong the contraction which takes place at the region stimulated may be very marked, resulting in the formation of strong folds in this region. The contraction may include the entire anterior end of the Amœba. Such a contraction destroys the attachment to the substratum, and the new pseudopodium now bursts out in some part of what was the posterior end of the body. The new course followed may then be at right angles to the old one, or at any greater angle, or the course may be exactly reversed, the new pseudopodium being formed at the posterior end. If the posterior end was much wrinkled or bore a pronounced roughened "tail," it is to be noticed that the new pseudopodium does not flow out directly from this, but to one side of it or above it (Fig. 67). Then as the Amœba moves in the reverse direction the body passes the old "tail," which finally brings up the rear again, fusing with the rough area produced by contraction of the region stimulated (Fig. 67, *b*). Of course, the new pseudopodium formed must come in contact with the substratum and become attached to it before locomotion in the new direction can occur. Sometimes the new pseudopodium formed is sent directly upward into the water; then there is no locomotion until the Amœba topples over, bringing the new pseudopodium in contact with the substratum.

At times when the anterior end is stimulated, two new pseudopodia are sent out in opposite directions on each side of the region stimulated.

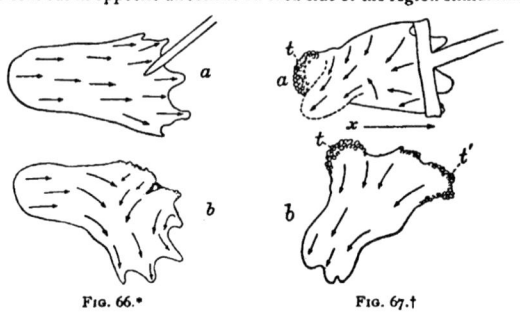

FIG. 66.*   FIG. 67.†

Both evidently pull on the Amœba, which becomes drawn out to form a narrow isthmus between them. Finally one end pulls the other away from its attachment to the bottom; the latter then contracts, and locomotion continues in the direction of the prevailing pseudopodium.

There is at times a peculiar additional feature of the reaction to

---

* FIG. 66.—Negative reaction to a mechanical stimulus in Amœba. An Amœba advancing in the direction shown by the arrows is stimulated strongly with the glass rod at the anterior end (at *a*). Thereupon the currents are changed and a new pseudopodium sent out as at *b*.

† FIG. 67.—Negative reaction to a mechanical stimulus when the anterior end is strongly stimulated. The arrow, *x*, shows the original direction of motion; the arrows in *a* show the currents immediately after the stimulation. A large pseudopodium is sent out from above and to one side of the former tail (*t*), as is shown by the broken outline. In *b* this pseudopodium has come in contact with the bottom; the arrows show the direction of the currents and of locomotion at this time; *t*, the original tail; *t'*, the new tail formed by the contraction of the anterior end.

strong stimuli. In some cases there is for an instant after a strong stimulation at the anterior end a sudden rush of protoplasm *toward* the region stimulated; this is immediately followed by the stoppage and contraction above described. Apparently this sudden rush toward the point stimulated is produced as follows: The first effect of the additional contraction caused by the stimulus is to release a certain amount of surface at the posterior edge of the attached area from its attachment to the substratum. This portion was nearly ready to become released in the ordinary course of events, so that probably a very slight shock would release it at once. Now, as I have shown on p. 167, when a portion of the lower surface of the Amœba is suddenly released from the substratum, it contracts, causing a strong forward current. This is what happens in the case under consideration. Later this current is stopped by the effect of the stimulus in the anterior region.

The surface currents in the reaction are changed in the same way as are the internal currents, and are throughout congruent with them. Particles moving forward on the upper surface of the Amœba stop after the stimulus, then move in the direction of the new forward current. This has been illustrated in detail for *Amœba verrucosa* (p. 143), so that we need not go into the matter further here.

The essential features of the negative reaction to a mechanical stimulus are, then, a contraction of the region stimulated, with the formation of a new pseudopodium in what may be considered the region of least resistance, followed by a change in the direction of the currents of protoplasm, thus altering the course of the Amœba.

By repeated mechanical stimuli it is possible to drive the Amœba in any desired direction. I have at times made use of this possibility in order to bring into contact two Amœbæ or two pieces of an Amœba whose courses lay in different directions. Such driving of an Amœba requires considerable skill and a rather high tension on the part of the operator. The new pseudopodium formed is stimulated to withdraw as often as it is formed, until it finally starts out in the desired direction. If it were possible to stimulate all of one side of an Amœba at once it would, of course, be driven directly toward the opposite side, even though the stimulus were weak. With chemical and some other stimuli, as we shall see, this is possible. With mechanical stimuli it is usually possible only when the stimulus is very strong. By drawing the tip of the glass rod along one side of a moving Amœba, it is often possible to make it flow directly toward the opposite side, as illustrated in Fig. 68. This point is important for an understanding of the effects of such stimuli as chemicals, heat, and light.

When a single pseudopodium is stimulated, it is merely withdrawn, wrinkling and becoming warty in the usual way; there may be no other effect on the movement of the animal.

Among the sweeping statements that one finds current in regard to the behavior of these low organisms is one to the effect that the Protist does not avoid an obstacle in its path. This statement is made for example by Ziehen, in his excellent *Leitfaden der physiologischen Psychologie*.* It is worth while, therefore, to describe in connection with the reactions to mechanical stimuli just how Amœba avoids an obstacle. Let us take a concrete case. An Amœba creeping with a broad, flat anterior end came in contact at the middle of its anterior edge with the end of a long filament of some sort (Fig. 69). The particular spot touched (*c*) ceased to move forward, becoming entirely quiet (reaction to a weak mechanical stimulus). On each side of it motion continued as before, so that after a time the filament projected into a notch in the middle of the anterior edge. Then gradually the

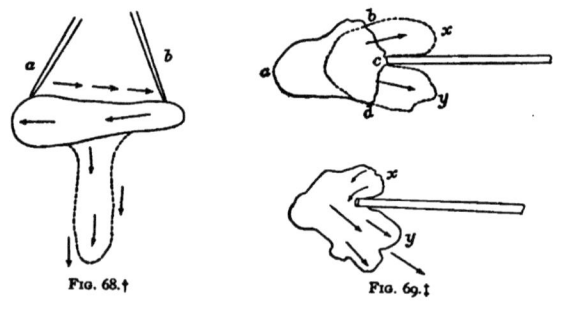

FIG. 68.†   FIG. 69.‡

forward movement ceased on the side *x* and increased at *y*, the pseudopodium *x* contracted, and its endosarc passed into *y*. The animal then continued its course in the direction indicated by *y*. It had thus changed its path so as to avoid the obstacle presented by the filament. Such cases are often seen.

---

* " Hindernissen weichen dieselben nicht aus " (*l. c.*, p. 11).

† FIG. 68.—An Amœba moving in the direction shown by the arrows in the unbroken outline is stimulated by drawing the tip of a glass rod along one side, from *a* to *b*. Thereupon a pseudopodium bursts out of the opposite side, as shown by the broken outline, and the Amœba continues locomotion in the direction so indicated.

‡ FIG. 69.—Method by which Amœba avoids an obstacle. The Amœba *a-b-c-d* comes in contact at *c* with the end of a filament. Thereupon motion at *c* ceases, while elsewhere it continues, so that after a time the Amœba has the position shown by the broken outline. Then the currents become changed in *x;* its substance passes into the pseudopodium *y*, and the Amœba continues to move in the direction indicated by the arrows in the lower figure.

Much more striking cases of the regulation of the movement in accordance with the position and changes in position of outward things ("automatic acts," Ziehen, *l. c.*) than are found in such a reaction, or even than in the possibility of driving an Amœba, will be described in the account of the food reactions (p. 196).

REACTIONS TO CHEMICAL STIMULI.

By analogy with the effects of mechanical stimuli we might expect to find a positive reaction to chemical stimuli. Such reactions doubtless occur, but I have not been able to demonstrate them under experimental conditions, and, so far as I am aware, no one else has succeeded in doing this. Stahl (1884) has shown the corresponding reaction to take place in the myxomycete plasmodium. Verworn (1890, p. 456) records an observation which he refers with much probability to a positive reaction to a chemical stimulus in Difflugia. If two conjugating Difflugias were separated, they crept directly together again, and it is difficult to see how the movements could have been directed save by some chemical. But I believe there is no instance of positive chemotaxis in Rhizopoda where the nature of the active substance is known and the reaction was controlled experimentally. A number of striking positive reactions, which should probably be attributed partly to chemical stimuli, are described later in connection with the attempts of Amœba to obtain food (p. 196).

The same state of affairs has existed hitherto with regard to our knowledge of a negative reaction to chemicals. In Amœba it is, however, not difficult to produce such reactions experimentally.

For this purpose only a small amount of the chemical must be used, so that it can act on but a limited portion of the body of the animal. If there is a considerable amount of the solution, diffusing over a large area, it reaches a strength sufficient to cause a reaction at about the same time over the whole body of the Amœba; thus the reaction is a general one, not involving movement in a definite direction. To produce a directed reaction there must be a decided difference in the strength of the solution on two sides of the organism.

The easiest method of producing the reaction, and the one giving at the same time the most striking results, is to dip the moistened tip of a capillary glass rod into powdered methyl green or methyline blue; then to bring this near one side or end of the Amœba, in an open drop of water. The chemical diffuses in a colored cloud; the reaction takes place when the edge of this cloud comes in contact with the Amœba.

The reaction is essentially the same as that to mechanical stimuli. The region stimulated stops suddenly, and about a second later contracts, while a current moves away from the side stimulated. This may meet the previously existing current coming from the original posterior end; the two turn to one side, and a pseudopodium starts out in a new place. If the stimulation took place at the anterior end and was limited to a small area, the new pseudopodium starts out at one side of the original anterior end; the new course followed, therefore, forms only a slight angle with the former one (Fig. 71, *a*). But if the stimulus affects all of one side of the body, or a still greater portion of its area, pseudopodia are sent out on the opposite side; the Amœba then creeps directly away from the source of diffusion of the chemical. This gives a typical case of negative "chemotaxis," the longitudinal axis being in the line of diffusion of the ions or molecules, with the anterior end directed away from the source of diffusion (Fig. 70). It will be noted that the reaction is exactly the same as that produced by mechanical stimuli; in "chemotaxis," where the animal is "oriented," we have the same process as in "driving" the Amœba in a definite direction by means of mechanical stimuli. All movement toward the chemical is inhibited, because this brings the protoplasm into a region where it is stimulated. Pseudopodia can be sent out, therefore, only on the side away from the chemical, and movement can occur only in that direction.

The surface currents are changed exactly as are the internal currents; the facts here are identical with those described for mechanical stimuli (p. 185). The surface currents are thus away from a chemical which causes a negative reaction (see Fig. 42, p. 144).

FIG. 70.*

A number of variations in the reactions to chemicals are shown in Fig. 71, all of them taken from actual experiments. As the figure shows, after stimulation frequently two pseudopodia start out in opposite directions, one finally prevailing over the other (Fig. 71, *d*).

The contraction due to the chemical is often very marked, the ectosarc against which the chemical impinges shrinking sharply together and becoming covered with folds (Fig. 71, *b*). With methyl green as the stimulus, the surface touched by the chemical is sometimes stained, so that the shrinkage in area is very precisely definable. With a solution of NaCl the shrinkage is extreme, while the opposite side spreads out widely, compensating, or more than compensating, for the decrease in surface caused by the shrinkage (Fig. 71, *b*).

The effect of substances not in the form of powder was tried in the following manner: A glass tube was drawn out to a very fine point, and into it was introduced some of the solution to be tested. The fine point of the tube was then brought close to the Amœba. Some of the

---

* FIG. 70.—Diagram of "negative chemotaxis" in Amœba. A chemical diffuses from a center, as indicated by the radii; the Amœba reacts in such a way as to creep directly away from the source of diffusion, in a line with the radii of diffusion.

chemical flowed slowly out, and its action on the Amœba could be observed. The results obtained by this method were very clear. A negative reaction, as described above, was observed in this way for solutions of the following substances: Methyl green, methyline blue, sodium chloride, potassium nitrate, potassium hydroxide, sodium carbonate, acetic acid, hydrochloric acid, cane sugar. Of these substances relatively strong solutions were used (usually about 1 per cent). Of course, the solution which came in contact with the Amœba was much weaker than this, being diluted by the surrounding water. Emphasis was not laid on the quantitative aspect of the matter; the question proposed was, How does the animal react? and not, How much is required to produce the reaction? Therefore, different strengths were employed

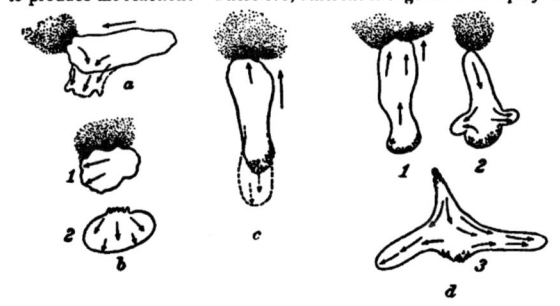

FIG. 71.*

till one was found that was effective. In any case, I do not know of any way in which one could determine the exact strength of the solution which comes in contact with the surface of the Amœba.

* FIG. 71.—Variations in reactions of Amœba to chemicals. The dotted area represents in each case the diffusing chemical. The arrows show the direction of the protoplasmic currents.

a. The chemical (methyl green) diffuses against the anterior end of an advancing Amœba; the latter reacts by sending out a new pseudopodium at one side of the anterior end and moving in the direction so indicated.

b. A solution of NaCl diffuses against the right side of a moving Amœba (1). The side affected contracts and wrinkles strongly, while the opposite side expands (2), the currents flowing in the direction indicated by the arrows.

c. A solution of NaCl diffuses against the anterior end of an advancing Amœba. The course is thereupon reversed, a broad pseudopodium, shown by the dotted line, pushing out from the upper surface of the posterior end above the tail.

d. A solution of methyline blue diffuses against the anterior end of an advancing Amœba (1); thereupon a pseudopodium is sent out on each side of the posterior end at right angles with the original course (2). Into these pseudopodia are drawn the body and the tail (3).

As a control experiment, distilled water was used in the tube in place of a chemical in solution. Amœba was found to react negatively to this also, though the reaction was less marked than with most of the chemicals. But this result, of course, rendered the experiments with solutions of chemicals in the tube indecisive, as the Amœba may have reacted to the distilled water in which the solutions were made up. The solutions were, therefore, made up with culture water, and the same results were obtained as before.

The results with the chemicals show merely that Amœba responds negatively to almost any solution differing markedly from that in which the animal is immersed, the precise chemical composition of the solution being of little consequence. The animal responded negatively not only to distilled water and to the chemicals mentioned, but also to tap water, and to water taken from other cultures than that in which the specimens occurred.

REACTIONS TO HEAT.

Verworn (1889, pp. 64–67) studied the directive influence of heat on the locomotion of Amœba by concentrating the sunlight on a small portion of the slide and leaving the rest dark, then observing the behavior of the Amœba on coming to the boundary of this lighted and heated area. The effects of the light proper were excluded by control experiments. It was found that on coming to the heated area Amœba remained quiet a moment, then contracted on the heated side, and sent out a pseudopodium on the opposite side. It then crept away in the direction indicated by this pseudopodium ("negative thermotropism").

My experiments differed from those of Verworn in employing conducted heat in place of radiant heat; thus there was no possibility of a complication from the effects of light. As Verworn sets forth, it is difficult to warm only one side of so small an object as an Amœba. I succeeded, however, in doing this in a very simple manner. For each experiment an Amœba was selected that was creeping on the under surface of the cover glass. This was placed in focus under an objective of a considerable focal distance, yet of high enough power so that the internal movements could be seen. A needle was then heated in a flame and its point was brought against the cover glass a short distance in advance of the Amœba. Control experiments had shown that the use of a needle at room temperature had no effect.

If the heated needle was placed at a proper distance from the Amœba, the phenomena follow as described by Verworn (*l. c.*). There was a short pause, then the side next to the needle contracted. A current of protoplasm passed toward the opposite side, at times meeting the current already in existence. A new pseudopodium was sent out, either on the side opposite the needle, or, in many cases, in a direction intermediate between this and the original one. The phenomena are in all respects identical with those seen in the negative reaction to mechanical stimuli. If the needle is brought a little nearer, so that the heat acts more strongly, there is a sudden strong contraction of the side affected. Simultaneously with this there is often, as in the case of strong mechanical stimuli, a sudden rush of the internal fluid *toward* the side stimulated. This lasts but an instant and is succeeded by a current away from the stimulated side, the formation of a pseudopodium on the unstimulated side, and locomotion in that direction. The sudden rush of internal contents toward the side affected is, I think, clearly due to the cause suggested under mechanical stimuli. Part of the posterior portion of the attached area of the Amœba is loosened from the substratum by the sudden contraction at the front end; this portion, therefore, contracts quickly and sends a current forward, as described on p. 168.

When the heat is still more powerful the entire Amœba is affected. It contracts and at the same time loses its attachment to the substratum. There is a strong momentary rush of the internal fluid toward the end which had been anterior, due to the cause set forth in the preceding paragraph. This ceases and the body becomes very irregular and ceases to move.

The reaction to local stimulation by heat is thus of essentially the same character as the reaction to mechanical stimuli and to chemicals.

Like Verworn (1889, p. 67), I have been unable to obtain a reaction to cold in Amœba.

REACTIONS TO OTHER SIMPLE STIMULI.

The reactions of Amœba to electricity and to light have been thoroughly studied by other authors, so that it will not be necessary to treat them in detail here. Only certain especially important points will be touched upon.

The reactions of Amœba to the continuous electric current have been studied in detail by Verworn (1890, *a;* 1897). I have repeated the experiments in order to determine by observation the direction of the surface currents of protoplasm during the reaction. For this purpose soot was mingled with the water containing the Amœbæ, and the electric current was passed through the preparation. The typical reaction as described by Verworn was observed in many cases, but the surface currents, of course, cannot be seen unless soot is resting upon or is attached to the surface of the animal, which happens only rarely. Finally a specimen of *Amœba proteus* was observed with a string of soot particles attached to one side (Fig. 72). The electric current was then passed through the preparation in such a way that the side bearing the soot was next to the anode (Fig. 72, *a*). The Amœba thereupon turned and began to move toward the cathode (*b*), dragging the particles of soot behind it for a short distance. Then the string of soot began to pass forward on the upper surface, in the usual way (*c*, *d*). This continued until the soot reached the anterior edge and dropped off the surface of the Amœba (*e*). The currents on the upper surface of Amœba are, then, forward (toward the cathode) in the reaction to the electric current, as well as in other cases.

On reversing the current the specimen described above began to move in the opposite direction toward the new cathode. In this and many other observed cases of the reversal of movement under the influence of the electric current, the reversal occurred in the same manner as when induced by other stimuli (see p. 183). That is, the new pseudopodium was sent out from one side of the attached (anterior) half of the body, changing the course a certain amount. From this new portion another new pseudopodium was sent out on the side toward the anode,

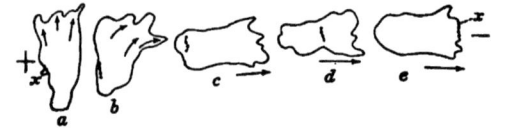

FIG. 72.*

and this continued until the direction of movement had been, by a gradual process, completely reversed. Verworn (*l. c.*) describes cases in which the reversal takes place suddenly, the new pseudopodium appearing at the original posterior end. This happens also at times, as we have seen, in the reactions to other stimuli (p. 183). It is to be noted that the reaction to the electric current is exactly that which would occur if the animal were strongly stimulated on the anode side.

Verworn (1889), Davenport (1897), and Harrington & Leaming (1900) have studied the reaction of Amœba to light. Verworn (*l. c.*, p. 97) found that light falling perpendicularly on one-half of the Amœba produced no reaction. Davenport (*l. c.*, pp. 186, 188) confirmed this result, but showed that when the light falls obliquely from one side on Amœba the animal reacts negatively. Harrington & Leaming (*l. c.*) found that when white light falls upon the Amœba from above the

---

* FIG. 72.—Movement of particles attached to the outer surface of Amœba in the reaction to the electric current. Anode and cathode are represented by the plus (+) and minus (—) signs. *a*, Form and direction of movement of the Amœba before the current is made; *x*, a chain of soot particles attached to one side; *b*, *c*, *d*, *e*, successive stages during the reaction. The chain of soot particles (*x*) passes to the upper surface and forward, reaching at *e* the anterior edge.

movements cease, while in red light they begin again; lights of other colors have various intermediate effects.

It seems to the writer that further experimentation is desirable on the results of a perpendicular illumination of one-half the animal. The difference between the results thus far obtained from such illumination and those from Davenport's experiments where light is admitted from one side is very remarkable. If this difference is constant, it is of much significance for the theory of light reactions. Possibly the lack of reaction when but one-half the animal is illuminated may be accounted for as follows: When one end of an Amœba is illuminated from below, as in Verworn's experiments, it is difficult or impossible to keep this difference of illumination constant for any considerable period. If the Amœba does not react at once it passes completely into the lighted area, where there is no cause for changing the direction of movement. On the other hand, in the case of light falling obliquely from one side, the different action of the light on the two sides is constant, so that in time a reaction is produced. The slowness of Amœba in reacting is such as to make this possibility worth considering. For further work from this point of view a source of powerful artificial light is needed. This I have not had at command during the present investigation.

It is evident that the reaction of Amœba to light falling from one side is exactly that which would be produced were the Amœba strongly stimulated on the side on which the light impinges.

For an account of the reactions of Amœba to general (not localized) stimuli, see Verworn, 1888, and the *Allgemeine Physiologie* of the same author.

SOME COMPLEX ACTIVITIES.

Under this heading I propose to describe certain striking phenomena in the behavior of Amœba, the stimuli to which are complex or not sharply definable. These concern the reactions of Amœba to food and to injuries, and the relations of one Amœba to another.

ACTIVITIES CONNECTED WITH FOOD-TAKING.

The behavior of Amœba in taking food or in attempting to take food shows many features of great interest for one attempting to understand the behavior of these organisms. I have observed the process of food-taking many times, and will describe it, together with a number of related activities.

Let us take a concrete case. A specimen of *Amœba proteus* was creeping about on a slide which contained many spherical cysts of *Euglena viridis*. One of these, which was not attached to the bottom (as most of them are), was lying in the path of the Amœba. The latter in its forward movement came against the cyst and pushed it forward a short distance. There was no evidence of a tendency of the cyst to adhere to the Amœba; on the contrary, it was pushed ahead as fast as the Amœba moved. The Amœba now put out a pseudopodium on each side of the cyst, while that part of the protoplasm immediately behind it stopped moving. Thus the cyst was enclosed in a little bay. On bringing the upper surface of the cyst into focus it could be seen that a thin layer of protoplasm was also sent over the cyst. The two pseudopodia enclosing the cyst now bent over at their free ends, so that the cyst could not be pushed away by movement of the Amœba. The two free ends finally met, leaving only a sort of transparent seam to show the place of contact. Later this disappeared, and the two pseudopodia fused completely. At the end of two minutes from the time that the Amœba first came in contact with it the cyst was completely enclosed. The Amœba now remained perfectly quiet for one minute, then crept away, carrying the cyst with it. With the cyst had been taken in some water, so that it was enclosed in a vacuole a little larger than itself. The walls of the vacuole had exactly the appearance of the ectosarc on the outer surface of the Amœba.

This is essentially the method of food taking that I have observed in a large number of cases in *Amœba proteus* and its relatives. The essential points are the sending out of pseudopodia on each side of and above the food body and the fusion of these pseudopodia at their free ends or edges, thus enclosing the food. In no case, in these species, was there any evidence that the Amœba was aided by the adherence of the food body to its protoplasm. On the contrary, there was a decided tendency for the food body to be pushed away, and an essential part of the process is the overcoming of this mechanical difficulty by sending out a pseudopodium on each side of the body and bending the ends of them together, so as to prevent slipping on the part of the food.* That this difficulty is no imaginary one will be shown later, in the description of cases where the Amœba was unable, after many efforts, to enclose the food.

It is commonly said that the posterior rough, tail-like portion of the body is especially important in the taking of food, though it is sometimes added that one rather more often sees the partly ingested food given out again in this region (see Leidy, 1879, p. 45; Penard, 1890, p. 81; 1902, p. 16).† I have never seen food taken in at this part of the body, though, as noted above, I have many times seen it taken at the anterior end. While it may be true that food is at times taken at the posterior end, I

*This lack of adherence between the protoplasm and the food substance is emphasized by Le Dantec (1894, p. 68) as a result of his careful studies on food-taking in Amœba.

† The references to food-taking at the posterior end seem all to go back to a paper by P. M. Duncan, "Studies amongst Amœbæ," in the Popular Science Review for 1877. I regret that I have been unable to see this paper.

believe that the supposed prevalence of this method of food-taking and of the giving off of incompletely ingested food here are really due to incorrect interpretation of another very common process. In its locomotion Amœba frequently comes in contact with diatoms, desmids, encysted Protozoa, etc. These it usually creeps over, so that they lie beneath it. As the Amœba progresses the objects come in contact with the posterior portion of the body, which is raised from the bottom and covered with a viscid secretion. Owing to this viscid substance the objects often cling to the under surface of this part of the body and are carried along with it. If observed at this time one cannot tell whether these objects have been ingested or not. But as a result of the method of movement of the Amœba they gradually pass to the posterior end, and are usually finally left behind. When such an object separates from the Amœba, becoming detached from its under surface, it appears exactly as if it were given off from within; it is only by observing the whole process from beginning to end that one can be sure of its exact nature. I am convinced that many of the supposed cases of the ingestion of food and of the ejection of food previously ingested at the posterior end are to be explained in this way. In all the detailed descriptions of food-taking in forms related to *Amœba proteus* that I have found in literature, the food was taken at the anterior end in a way similar to that which I have described above (see Carter, 1863, p. 45; Wallich, 1863, c, p. 453; Leidy, 1879, p. 49; Le Dantec, 1894, p. 68; also Bütschli, 1880, p. 117).

In *Amœba verrucosa* and the other species which do not often form pseudopodia food is taken in a somewhat different manner. Food-taking in *Amœba verrucosa* has been described by Rhumbler (1898, p. 205). Penard (1902), though he spent much time studying this species, says that he has not observed the taking of food, and thinks it must occur only rarely. In my own cultures specimens of this species taking food were positively abundant. I have seen the process much oftener than in other species. In *A. verrucosa* and its relatives food-taking is greatly aided by the tendency of foreign particles to cling to the surface of the body, a tendency which we found so convenient for determining the movement of points on the body surface (see pp. 140–146). This adhesiveness of the outer surface compensates for the lack of formation of pseudopodia in these species. The outer surface gradually sinks in at the point where the food body is attached to it. The latter is thus carried to the inside of the body, surrounded by a pouch of ectosarc. This pouch becomes separated from the outer ectosarc. The food is thus completely enveloped and later digested. Not only large objects, but often very small ones, spores of algæ, small diatoms, flagellates, etc., are taken in in this way. Rhumbler (*l. c.*, p. 208) has given a very interesting account of the rolling up and ingestion of threads of Oscillaria by this species (see p. 223).*

PURSUIT OF FOOD.

Amœba proteus does not always succeed in ingesting its food so easily as in the case just described (p. 194). There is, as noted above, a tendency for the food body to be pushed away by the forward movement at the anterior end of the Amœba, and this sometimes gives serious difficulty. In such cases Amœba may show what would be called in higher organisms remarkable pertinacity in continuing its

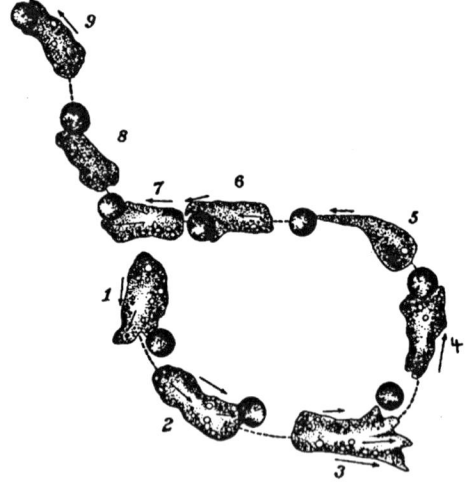

FIG. 73.†

attempts to ingest the food. This will be illustrated from a concrete case (Fig. 73).

An *Amœba proteus* was creeping toward an encysted Euglena. The latter was perfectly spherical and very easily moved, so that when the anterior edge of the Amœba came in contact with it the cyst merely moved forward a little and slipped to one side (the left). The Amœba

---

* Leidy (1879, p. 86) gives a very similar account of the ingestion of filaments of algæ in Dinamœba.

† FIG. 73.—Amœba following a rolling Euglena cyst. Nos. 1–9 show successive positions occupied by Amœba and cyst. See text for explanation.

thereupon altered its course so as to follow the cyst (Fig. 73, 1). The cyst was shoved forward again and again, a little to the left; the Amœba continued to follow. This continued until the two had traversed about one-fourth the circumference of a circle; then (at 3) the cyst, when pushed forward, rolled to the left quite out of contact with the Amœba. The latter then continued forward with broad anterior edge in a direction which would have taken it past the cyst. But a small pseudopodium on its left side came in contact with the cyst. The Amœba thereupon turned again and followed the rolling cyst. At times it sent out two pseudopodia, one on each side of the cyst (as at 4), as if trying to inclose the latter, but the ball-like cyst rolled so easily that this did not succeed. At other times a single very long, slender pseudopodium was sent out, only the tip of which remained in contact with the cyst (5). Then the body of the Amœba was brought up from the rear and the cyst pushed farther. This continued until the rolling cyst and the following Amœba had described almost a complete circle, returning nearly to the point where the Amœba had first come in contact with the cyst. At this point, owing to the form of the anterior end of the Amœba (7) the cyst rolled to the right instead of to the left as it was pushed forward. The Amœba followed (8, 9). This new path was continued for two or three times the length of the Amœba. The direction in which the ball was rolling would soon have brought it against an impediment, and I thought it possible that the Amœba might succeed in ingesting it after all. But at this point one of those troublesome disturbers of the peace in microscopic work, a ciliate infusorian, came near and whisked the ball away in its ciliary current. After the ball was carried away the Amœba continued to follow in the same direction for only a very short distance, about one-fifth its length, then reversed its course and went elsewhere.

The movements of Amœba are, of course, very slow, and the behavior described required a considerable period of time—10 or 15 minutes, I should judge. The whole scene made really an extraordinary impression on the observer, and it is difficult in describing it to refrain from the use of words that imply a great deal of resemblance between Amœba and immensely higher organisms. One seems to see that the Amœba is *trying* to obtain this cyst for food, that it puts forth *efforts* to accomplish this in various ways, and that it shows remarkable *pertinacity* in continuing its *attempts* to ingest the food when it meets with difficulty. Indeed, the scene could be described in a much more vivid and interesting way by the use of terms still more anthromorphic in tendency.

I have seen a large number of cases like that above described; in some of my cultures containing many specimens of *Amœba proteus* and many Euglena cysts it was not at all rare to find the animals engaged in thus following a rolling ball of food. I have made full notes and sketches of a number of other cases, but they show nothing different in principle from that above described, so that it is not worth while to enter into details. One further point is, however, worthy of special note. Often a single pseudopodium comes in contact with such a cyst and stretches out toward it, while the remainder of the Amœba continues on its course, away from the cyst. The pseudopodium in contact then stretches out as far as possible, keeping in contact with the cyst and often pushing it ahead (Fig. 74), until it is finally pulled bodily away by the movements of the whole Amœba. Apparently this one pseudopodium reacts to the stimulus quite independently of the remainder of the body. Again, two pseudopodia on opposite sides of the body may each come in contact with a cyst. Each then stretches

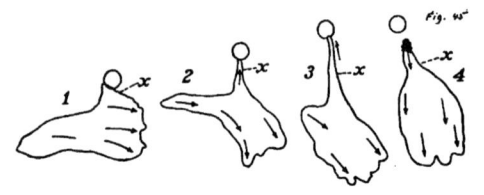

FIG. 74.*

out, pulling a portion of the body with it, and follows its cyst, until the body forms two lobes, connected only by a narrow isthmus. Finally, one half succeeds in pulling the other away from the attachment to the substratum, and the entire Amœba follows the victorious pseudopodium.

Mechanical stimuli are, of course, involved in the above reactions; perhaps also chemical stimuli from the cyst. It is important from the theoretical standpoint to note that the movement of particles on the surface of the Amœba is *toward* the object causing the reaction. This I have been so fortunate as to have opportunity of observing in several cases.

OTHER AMŒBÆ AS FOOD.

Amœbæ frequently prey upon each other, as Leidy has already described and figured (1879, p. 94; plate 7, Figs. 12-19). But the victim does not always conduct itself so passively as in the case described by Leidy, and sometimes finally escapes from its pursuer. A description

---

* FIG. 74.—A single pseudopodium (*x*) reacts positively to a Euglena cyst, the protoplasm flowing in direction of cyst and pushing it forward, while remainder of the Amœba moves in another direction; 1-4, successive forms taken. At 4 the reacting pseudopodium is pulled away from the cyst, and then contracts.

of two or three concrete cases among those which I have observed in *Amœba angulata* will bring out the nature of the behavior under such conditions. Penard (1902, p. 700) mentions that he has seen one Amœba pursue and finally capture another, but does not give a detailed account of the process.

(1) Two Amœbæ were observed, a large one and a small one, the former apparently attempting to swallow the latter (Fig. 75). The small Amœba was creeping rapidly forward, while its wrinkled posterior portion was enveloped by the anterior part of the larger Amœba. The large Amœba had the anterior portion of its body quite hollowed out, so as to form a cavity large enough to contain the entire small

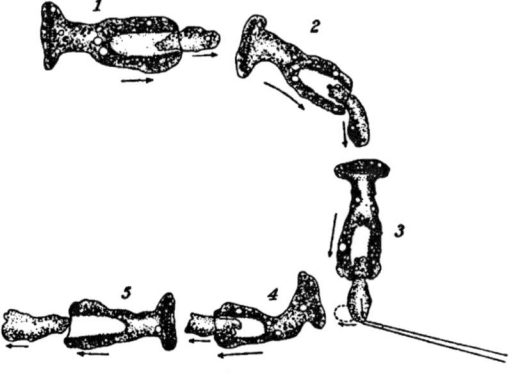

FIG. 75.*

Amœba, and in the anterior portion of this cavity was inclosed the hinder portion of the body of the smaller Amœba. Whether this cavity was bounded below as well as above and at the sides by protoplasm I could not determine with certainty. The large Amœba was following the small one, moving at about the same rate. There was no union between the protoplasm of the two; on the contrary the boundaries of both were clearly defined, and they seemed to be only slightly in contact, the posterior end of the small specimen moving easily within the cavity of the other. As they moved forward, sometimes the posterior specimen flowed a little faster, and then a little more of the smaller one became enveloped; at other times the smaller Amœba moved a little faster, and then withdrew a part of its inclosed

* FIG. 75.—Pursuit of one Amœba by another. See text for explanation.

tail. They progressed in this fashion for a long distance, many times their own length. I watched them thus for more than 10 minutes. The smaller, anterior specimen frequently altered its course; the posterior one followed. I stimulated the anterior end of the small specimen with the tip of a glass rod (Fig. 75, 3); it turned at a right angle, and the posterior specimen followed. After about 12 minutes it could be seen that the smaller specimen was moving slightly faster than the other and was slowly withdrawing its posterior end. Finally it pulled completely away from the large Amœba, which was still following as rapidly as possible. After the small Amœba had completely escaped the large one stopped and remained entirely quiet for a few seconds. The large cavity in its anterior portion, which it had prepared for the reception of the small Amœba, and which extended back behind the middle of the body, was still very evident. After a time the Amœba began to change form and sent out pseudopodia irregularly in all directions. The smaller Amœba continued its forward locomotion as long as observed. The performance is illustrated in Fig. 75, from sketches made while it was in progress.

(2) In a second case I was able to observe the beginning as well as the end of this microscopical drama (Fig. 76). I had attempted to cut an Amœba in two with the tip of a glass rod, in the manner described later. The posterior third of the Amœba, in the form of a wrinkled ball, remained attached to the body only by a slender cord, the remains of the ectosarc. The Amœba began to creep away, dragging with it this ball. I will call this Amœba *a*, while the ball will be designated *b*. A larger Amœba (*c*) approached, moving at right angles to the path of the first Amœba; its course accidentally brought it into contact with the ball *b*, which was dragging past its front. Amœba *c* thereupon turned, followed Amœba *a*, and began to engulf the ball *b*. A large cavity was formed in the anterior end of Amœba *c*, reaching back nearly or quite to its middle, and much more than sufficient to contain the ball *b*. Amœba *a* now turned into a new path; Amœba *c* followed (Fig. 76 at 4). After the pursuit had lasted for some time the ball *b* had become completely enveloped by Amœba *c;* the cord connecting it with Amœba *a* broke, and the latter went on its way (at 5) and disappears from our account. Now the anterior opening of the cavity in Amœba *c* became partly closed, leaving a slender canal (5). The ball *b* was thus completely inclosed, together with a quantity of water. There was no union or adhesion of the protoplasm of *b* and *c;* on the contrary (as the sequel will show clearly) both remained quite separate, *c* merely inclosing *b*.

Now the large Amœba *c* stopped, then began to move in another direction (Fig. 76, 5–6), carrying with it its meal. But the meal, the ball *b*, now began to show signs of life, sent out pseudopodia, and, indeed, became very active. We shall henceforth, therefore, speak of it as Amœba *b*. It began to creep out through the still open canal, sending forth its pseudopodia to the outside (Fig. 76, 7). Thereupon Amœba *c* sent forth its pseudopodia in the same direction, and after creeping in that direction several times its own length, again completely

FIG. 76.*

inclosed *b* (7-8). The latter again partly escaped (9), and was again engulfed completely (10). Amœba *c* now started again in the opposite direction (11), whereupon Amœba *b*, by a few rapid movements, escaped entirely from the posterior end of *c*, and was free, being completely separated from *c* (11-12). Thereupon *c* reversed its course (12), crept up to *b*, engulfed it completely again (13), and started away. Amœba *b*

* FIG. 76.—Pursuit, capture, and ingestion of one Amœba by another; escape of captured Amœba and its recapture; final escape. See text for detailed account.

now contracted into a ball, its protoplasm clearly set off from the protoplasm of its captor, and remained quiet for a time. Apparently the drama was over. Amœba *c* went on its way for about five minutes, without any sign of life in *b*. In the movements of the Amœba *c* the ball *b* gradually became transferred to the posterior end of *c*, until finally there was only a thin layer between *b* and the outer water. Now *b* began to move again, sent out pseudopodia to the outside through the thin wall, and then passed bodily out into the water (14). This time Amœba *c* did not return and recapture *b*. The two Amœbæ moved in different directions and remained completely separated. The whole performance occupied, I should judge, about 12 to 15 minutes (the time was not taken till several minutes after the beginning).

After working with simple stimuli and getting always direct simple responses, so that one begins to feel that he understands the behavior of the animal, it is somewhat bewildering to become a spectator of so striking and complicated a drama. If we attempt an analysis of the observed behavior of the Amœba *c* into stimuli and reactions, we obtain some such a result as follows: At first the stimulus of contact with *b*, and perhaps a chemical stimulus from the same source, causes the Amœba *c* to react by flowing toward *b*, and at the same time to change form, so as to hollow out the anterior end. Later, every change in the direction of movement of *a* and *b* induces a corresponding change in the direction of movement of *c*; there is a finely co-ordinated adaptation of the latter to the movements of the former. After the separation of *b* from *a*, the movement of *c* (at 5–6) in a different direction may have been due to some external stimulus. But what is the stimulus for the change of direction of locomotion in the Amœba *c* at 7 when *b* has begun to escape? And why does Amœba *c* go in that direction only long enough to get *b* firmly inclosed again, then reverse its course? And, finally, why does Amœba *c* reverse its course at 11–13, when *b* has entirely escaped, and continue in this reversed direction till it reaches and recaptures *b*? The action is remarkably like that of a higher animal. Doubtless we must assume chemical and mechanical stimuli as directives for each of the movements of *c*, but the analysis so obtained seems not very complete or satisfactory.

REACTIONS TO INJURIES.

Certain cases that belong under the heading of reactions to injuries have already been described as evidences of the contractility of the ectosarc; for these page 180 should be consulted. The cases which we take up here are of a different character. They concern *Amœba angulata*.

Jensen (1896) has shown in the case of certain Foraminifera that two pieces of protoplasm from the same individual will readily unite if brought in contact, while pieces from different individuals will not thus unite. I was interested in the question as to whether this would hold true also for Amœba, and for that purpose undertook to cut specimens in two. With the fine tip of a glass rod it is possible, under the microscope, in the open drop, to cut in two an elongated Amœba at almost any desired point. The sharp point of the rod is simply drawn across the Amœba as it lies outspread on the substratum.

In this operation it was found that the Amœba was not, as a rule, at first completely cut in two by the stroke itself. The endosarc is divided completely, but the two halves are still connected by a thin layer of ectosarc, which resists the cutting, and shows fine longitudinal striations; these may be merely longitudinal folds (Fig. 77, 2). This thin layer of ectosarc seems very tenacious.

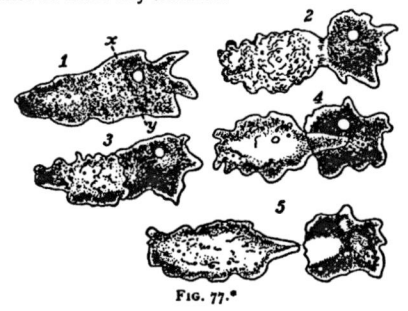

FIG. 77.*

The Amœba is thus left in the condition shown in Fig. 77, 2. The two halves usually both contract strongly. Now ensues a very peculiar process. One of the two halves begins to send out pseudopodia in such a way as to partly inclose the other (3); the second half is thus drawn as a narrow wedge-shaped mass inside of the other, as at 4. It seems to be usually the half that contains the nucleus that envelopes the other, though, as will be shown later, the nucleus is not necessary for this reaction. If the piece thus embraced is considerably smaller than the other, it may become completely inclosed, and is then carried away, appearing like a mass of food. It does not become fused with the remainder of the protoplasm, but there is a sharp boundary between it and that which envelopes it. Specimens were followed for 10 minutes after thus inclosing a piece of their own bodies; during this time no marked change was seen to occur in the inclosed piece.

In the much more common cases where the two pieces are nearly

---

*FIG. 77.—Reaction of Amœba to injury. See text.

equal in size, or when it is the smaller piece that begins to envelop the larger, the process results differently. After one piece has been drawn far into the other (Fig. 77, 4), both seem to contract strongly, whereupon the connecting band of ectosarc breaks, the partly inclosed piece is squeezed out of the other; and the two separate. Usually each retains its form for a few seconds after separation; one bearing a slender truncate pyramid or cone, while the other shows a deep depression corresponding to this pyramid (Fig. 77, 5). After a time both halves change form and move away. Usually the half which partly inclosed the other becomes active long before the other, but this is not invariably true.

In a large number of cases observed it was the part which contained the nucleus that attempted to envelop the other half. In order to determine whether the nucleus plays a necessary part in this performance, I tried the following experiment: After the half which had no nucleus had again become active and was moving about, I cut it in two, as before. Now one half of this piece partly enveloped the other in the usual way, thus showing that the nucleus is not necessary for this reaction.

These results should be compared with Penard's observations on injured specimens of *A. terricola*, noted on page 180 of the present paper.

As to the question which these experiments were originally intended to answer, whether two pieces of a single Amœba would reunite after separation, my results were negative. After the two pieces had begun to move about freely they were induced to come in contact, or sometimes they came in contact accidentally; but in no case was there any union. Prowazek (1901, p. 93) obtained the same result with small species of Amœba, but in a larger undetermined species he succeeded in bringing about a union of pieces not only from the same individual, but from different individuals.

PHYSICAL THEORIES AND PHYSICAL IMITATIONS OF AMŒBOID MOVEMENTS.

THE SURFACE TENSION THEORY.

The movements of Amœba as presented by Bütschli (1880, 1892) and Rhumbler (1898) (see Figs. 30–33) are exactly those of a drop of fluid moving as a result of a local change in surface tension (Fig. 34). It was, therefore, natural to assume that the cause of the movements is the same in the two cases. This is the view taken by the two authors named. According to Bütschli, Rhumbler, and many other authors, Amœba is a drop of complex fluid which moves about as a result of local changes in surface tension.

In the foregoing investigation it has been shown that the movements in Amœba are not of the character supposed by Bütschli and Rhumbler. There is, indeed, very little resemblance between the movements of Amœba and those of an inorganic drop moving as a result of a local change in surface tension. The difference is clearly brought out by a comparison of Fig. 58, showing the currents in Amœba, with Fig. 34, showing those in the inorganic drop. The more striking differences are as follows:

(1) In the drop moving as a result of a local change in surface tension the currents on the surface are (and must be) *away from* the side on which a projection is formed and toward which the drop is moving; in the Amœba the surface current is *toward* this side.

(2) In the drop the surface currents are in a direction opposed to that of the axial current; in Amœba surface currents and axial current are in the same direction.

(3) The movement of Amœba is in the nature of rolling, the upper surface passing continually around the anterior end and becoming the lower surface. In the inorganic drop there is no such rolling movement, but the interior portions of the fluid are continually passing to the surface at the anterior end.

Clearly, then, the forces producing the movements in the two cases are not acting in the same manner. The locomotion of Amœba is demonstrably *not* due to a local decrease in surface tension on the side toward which the animal is moving.

This becomes still clearer when we consider in detail the method by which the movements are produced in a drop of inorganic fluid as a result of a local decrease in its surface tension.

The phenomena of surface tension are usually considered to be the result of the uncompensated attractions of those particles of the fluid which are near to the surface. Such particles are attracted inward and laterally, but not outward (or less strongly outward). The resulting forces may be considered as resolvable into two components, one acting tangent to the surface, the other acting perpendicular to the surface. The former is what may be called *surface tension proper;* the latter is often spoken of as *normal pressure.* The result is very much as if the fluid were covered with a stretched India rubber membrane. These relations are well set forth in a recent paper of Jensen (1901). It is further to be noted that these two components, surface tension and normal pressure, are two aspects of one and the same thing, and, therefore, vary together and from the same causes; they can not be separated either theoretically or experimentally. Whenever one of these factors increases or decreases, the other shows a corresponding change. The two are often spoken of (together with the pressure due to curvature of the surface) as *surface tension* (see Jensen, *l. c.*).

Now, when the interattraction of the particles at a certain region of the surface of a mass of fluid is decreased, the pressure inward and the tension along the surface are decreased. As the pressure remains the same elsewhere, fluid tends to be pressed out at the point where the pressure is lowered; thus a projection may be formed here. As the tension remains the same elsewhere, the remainder of the surface of the drop pulls harder than that of the region under consideration; hence it pulls the surface of the fluid away from the region where the tension is lowered. The effect is similar to that which would be produced if one portion of a stretched sheet of India rubber were weakened or cut; the remainder of the sheet would pull away from this region. Thus there are produced the currents characteristic of such a drop of fluid—an axial current toward the region of lowered tension, surface currents away from the region of lowered tension (Fig. 34). An increase in the tension at the opposite side would produce exactly the same currents, as Rhumbler (1898, p. 188) has set forth, the axial current being always toward the region of lowest tension, the surface currents in the opposite direction.

In the moving Amœba, as we have seen, the currents are by no means of this character. The axial current and the surface current are congruent, and both are in the direction of locomotion. Such movement could not be produced by a local decrease in the surface tension of some part of the body surface.

The formation of pseudopodia is, as we have seen, essentially the same process as the forward movement at the anterior end of the Amœba. On the upper surface of a pseudopodium that is in contact with the substratum there is a forward movement, so that particles clinging to the upper surface are carried over the tip; the currents which must result from a decrease in surface tension are not present. On the contrary, there is a current on the surface in the opposite direction from that required. The formation of such a pseudopodium can not, then, be due to a local decrease in surface tension.

The same is true, essentially, when a pseudopodium is sent out into the water, not coming in contact with a surface. In such a case, as we have seen, the entire surface moves outward, in the same direction as the tip; there is no such backward movement as the theory requires.

Altogether, it is clear that the supposed resemblance between the movements and internal currents of Amœba and those of a drop of fluid moving as a result of a local increase or decrease of surface tension does not exist. We must conclude that the movements of Amœba are not due to local changes in surface tension.

One might be tempted to inquire whether the movement of Amœba could not be explained by considering separately the action of the two factors in surface tension, the "surface tension proper" and the "normal pressure." If the normal pressure, directed inward, were decreased in a certain region, while the tangential factor, the "surface tension proper," were not decreased, were perhaps even increased, could not pseudopodia be formed as actually occurs, without any backward current on the surface? Jensen seems to lean toward the possibility of such action when he speaks of the variation of one of these factors without the other (Jensen 1901, p. 374).*

But with such an inquiry should we not leave the field of realities to wander among abstractions? One who is not a physicist can, of course, not speak positively on such a point. Yet, so far as I am able to discover from the results of experiments and from the theories of surface tension, the state of the case is about as follows: The tangential tension and the normal pressure are not two different things; they are only different aspects of one and the same thing. Viewed from the standpoint of "energetics," what the physical experiments show liquids to possess is surface energy, in virtue of which they tend to decrease their surface and resist an increase of surface, however these changes are brought about (see Ostwald, 1902, p. 197). When the "surface tension is decreased" in a certain region (x) of a fluid mass, this signifies that the tendency to a decrease of surface and the resistance to an increase of surface is lessened in this region. As a result, the remainder (y) of the fluid decreases its surface at the expense of the region x; the latter is thus compelled to increase its surface. This takes place by simultaneous passage of the contents of y into x and of the surface of x on to y, the two operations being essentially one, and both having the result of decreasing the surface of y and increasing that of x. There would seem to be no ground, theoretical or experimental, for supposing that in a fluid one of these operations could take place without the other. An attempted explanation of this sort would be, if these considerations are correct, not a physical explanation, but a purely hypothetical one, working with conditions not known to exist. The whole value of the surface tension theory lies in its direct reference back to the results of physical experiments—in its fidelity to the results of such experiments. As soon as it leaves this ground it becomes of no more value than the thousand and one other hypotheses that have been constructed for the explanation of contractility.

Further, even this purely hypothetical explanation could not account for the forward currents on the upper surface of Amœba, nor for the transference of portions of the body surface to the surface of a pseudopodium. In any form we can give it, the theory that the movement is due to local changes in surface tension is not in agreement with the observed phenomena.

* Though elsewhere he speaks of the necessity of their varying together.

BERTHOLD'S THEORY THAT ONE-SIDED ADHERENCE TO THE SUB-
STRATUM IS THE CAUSE OF LOCOMOTION.

If, then, the movements of Amœba are not due to local decrease in surface tension, with the formation of "Ausbreitungscentren" (Bütschli), is it possible to find a physical explanation for them? In taking up this question we must consider separately (1) locomotion, and (2) the formation of pseudopodia.

Observation and experiment indicate, as we have seen in the observational portion of this paper, that Amœba is a drop of fluid which becomes attached to the substratum in front and pulls itself forward, the pull extending backward from the attached region over the upper surface, and producing a rolling motion.

Now, *a drop of inorganic fluid under the influence of similar forces moves in precisely the same manner.* There is no great difficulty in causing a drop of inorganic fluid to adhere more strongly to the substratum on one side than elsewhere. When this is brought about the drop moves toward the more adherent side by a rolling motion, precisely like that of Amœba. By a proper arrangement of the conditions almost every detail of amœboid locomotion may be closely imitated.

That this is the method of movement in Amœba was the theory maintained by Berthold (1886), though it is rather curious that the supposed facts on which he based this view were incorrect. Berthold confirmed on the basis of observations on *Amœba verrucosa* (!) and other species the account of the currents in Amœba given by Schulze (see p. 137 and Fig. 37); that is, such currents as would be consistent with the theory of local decrease in surface tension, but are quite inconsistent with his own theory. He rejected the theory that locomotion is due to a decrease in surface tension at the anterior end, on the ground that no currents are to be observed in the surrounding water, as this theory demands. Berthold held that the locomotion is due to the spreading out of the anterior end of the fluid mass on the surface of a solid, this spreading out being due to adhesion between the fluid and the solid. Unfortunately for the understanding of his theory, he tried to bring this into relation with many other much less simple phenomena. In particular he compared the movements to those of a drop of water on a glass plate, which flees when a rod wet with ether is brought near one side. This was an unfortunate comparison, as the movements in a drop of water under such circumstances are of a character entirely different from those produced when a mass of fluid adheres by one side to a solid. The movement in a drop of water fleeing from the etherized rod is a result of the currents produced by the lowering of the surface tension on one side, as Bütschli (1892, pp. 191 and 194), has shown. This comparison, and Berthold's discussion of the relations between spreading out on the surface of solids and on the surface of liquids, together with his incorrect idea of the currents in such spreading out on a solid, have served to distract the attention of investigators from the really simple essential features of such a theory. Berthold did not attempt to study directly the currents and other movements of a drop of fluid moving as a result of one-sided adherence to a solid. We may, therefore, leave his account and examine for ourselves the phenomena in question.

EXPERIMENTAL IMITATION OF THE LOCOMOTION OF AMŒBA.

The experiments with inorganic fluids may be performed as follows: A piece of smooth cardboard, such as the Bristol board used for drawing, is placed on the level bottom of a shallow vessel, such as a Petrie dish, and soaked with bone oil by spreading the latter over its surface. A small area on the surface of the board is protected from the oil by placing upon it a drop of water. After the board has become well soaked, the drop of water is removed with a pipette, leaving this spot merely damp, while a layer of oil some millimeters deep is poured into the vessel, covering the cardboard completely. A drop of glycerine or of water is then introduced; this settles to the bottom, but adheres to it only slightly. A drop of glycerine is in some respects preferable, as its movements are slower. To the drop should be added beforehand a quantity of soot, in order to make its internal movements visible. Some of the soot remains on the surface, projecting out into the oil, thus making it possible to observe the surface currents.

If the drop is brought close to the spot on the cardboard that was protected from the oil, so that one side comes in contact with this region, the edge of the glycerine or water drop spreads out over this area. Thereupon the remainder of the drop is pulled in that direction, till the whole drop takes up its position over the protected spot. In the movement of the drop toward the area to which one side adheres, it rolls exactly as Amœba does. The currents on the upper surface and within the drop are forward. Toward the sides the currents are somewhat less marked, and on the under surface they cease entirely; particles within the drop but in contact with the lower surface are not moved at all. The forward current is most rapid in front, becoming slower at the rear, exactly as in Amœba. At the posterior end the surface rolls upward; particles on the surface which were at first on the bottom may be seen to pass upward around the posterior end and then forward, as in Amœba. The form of the drop may become much elongated; the anterior edge is thin, the posterior end thick and rounded. In all these respects the drop resembles the moving Amœba.

By inclining the vessel the drop may be made to roll away from the attractive spot; then when the level is restored it moves back again. By repeating this process the movements may be studied in detail. For studying the movements in all parts except at the anterior edge another more convenient method may be employed. A small piece of wood may be brought against one side of the drop; toward this it moves in the manner just described. If the piece of wood is moved continually in a certain direction, the drop follows, and its movements may be examined with ease. In this case the anterior edge, of course, is not thin and pressed against the surface, but otherwise the movements are the same.

By proper modifications further details of the movement of Amœba are exactly imitated. Thus a quantity of sand grains or other heavy objects may be added to the drop. In the movement these collect at the posterior end, as happens with the coarse internal contents in Amœba. A large, spherical bubble of oil may be introduced into the drop, in imitation of the contractile vacuole; this likewise stays near the posterior end. When a considerable quantity of heavy material is collected at the posterior end, the latter becomes drawn out into a sort of pouch, which is dragged along, its substance not partaking of the currents shown by the remainder of the drop. It thus plays the same part as the well-known posterior appendage of Amœba. Material passes up from the bottom to the upper surface on each side of this posterior pouch, just as happens in Amœba (see p. 168). Particles clinging to the outer surface at the posterior end are often dragged along for a considerable time, then finally pass upward to the upper surface and so forward, exactly as described for Amœba (p. 169).*

In another detail the movements of the drop of water or glycerine are strikingly like those of Amœba. As we have seen, the current is most rapid at the anterior end in both cases, becoming as a rule slow toward the rear. But I have pointed out that in Amœba the movements at the posterior end are not uniform. Sometimes the under surface remains attached to the bottom longer than usual, then, when it becomes detached over a considerable area at once, there is a sudden rush of the fluid forward from the posterior region (p. 168). Exactly the same thing is to be observed in the inorganic drop. The bottom is not uniform, so that sometimes the posterior end clings to it longer than usual; this end is then drawn out, and when it is finally released

---

* It may be worth while to state that these experiments on inorganic fluids were performed after the work on the movements of Amœba had been completed and the description entirely written in the form given in the preceding pages. No details of the movements of Amœba were added after the behavior of the inorganic drops had been studied.

there is a sudden rush forward of the internal fluid from this region, giving the movement a jerky character.

Still another resemblance in detail between the movements of the inorganic drop and of Amœba may be noted at times. As we have seen, in the posterior part of Amœba that is detached from the bottom there is a movement forward not only on the upper surface, but also a slow movement on the lower surface; the entire posterior region is contracting. The same thing may be seen in the inorganic drop. The phenomenon in question is not so regular here, because the posterior half usually still clings to the surface to a certain extent, while in Amœba it is as a rule entirely free. But when the posterior half of the inorganic drop does become entirely free, it is seen to contract as a whole, with a forward movement on both upper and lower surfaces, exactly as in Amœba.

One may even see at times, under special conditions, a slight turning backward of the current at the sides of the anterior end, such as has been described by a number of authors for Amœba (see p. 137). This occurs when the drop is slender and elongated, and the area on which it spreads out is broad. On coming in contact with the area, the end of the drop rushes forward and spreads out. If the whole width of the area is not covered at first, some of the particles that have moved forward curve outward and a little backward till the area is quite covered.

Altogether, the resemblance between the movements of the inorganic drop and those of Amœba is extraordinary, extending even to details. What are the forces at work in such a drop, and in how far may they be supposed to be active also in Amœba?

The spreading out of the drop of glycerine or water at the anterior end is due to its adherence here to the substratum. The remainder of the movements of the inorganic drop are due to the interplay of surface tension and adhesion to the substratum. As a result of surface tension the drop seeks to regain its spherical form; hence the posterior part is pulled forward, the force required to accomplish this being less than would be demanded for freeing the anterior edge from the substratum. In the pulling forward of the posterior portion the adherence of the lower surface to the bottom keeps this surface from moving; hence the upper surface moves forward while the lower surface remains quiet or moves forward only very slowly; the movement is thus converted into a rolling motion. The details given above depend merely upon the relative part played by adherence and surface tension, with the resistance offered by the weight and inertia of particles inclosed in the drop.

In Amœba, so far as the evidence of observation goes, the conditions are similar. Amœba adheres to the substratum and spreads out in a similar manner. In one respect there is, of course, a striking difference between the two cases. Amœba does not require that the substratum should be different on its two sides in order that there should be movement; on the contrary, it may move steadily in a certain direction on a uniform surface. The mechanism of the movement might, nevertheless, be the same as in the inorganic drop. Chemically different substances show different degrees of adherence to the same surface. It may be supposed, therefore, that there is a chemical difference between the anterior and posterior regions of Amœba of such a nature that the anterior region clings to the surface while the posterior region does not. This chemical difference must, of course, be continually renewed, since new parts of the body continually come in contact with the substratum.

It may, however, be questioned whether the adhesion of Amœba to the substratum is of the same character as the adhesion of a drop of water to glass; in other words, whether Amœba really plays here the part of a fluid, and "wets" the substratum. This was the view taken by Berthold (1886) and, if I understand him correctly, Le Dantec (1895). Apparently opposed to such a view is the fact that Amœba may creep on the under side of the surface film of water, as I have often observed. This surface film is, of course, fluid; if in adhesion Amœba itself also plays the part of a fluid, we should have two fluids in contact, having the same relation of attraction or adhesion that a fluid has for a solid that it "wets"; that is, the particles of each fluid have a greater attraction for those of the other fluid than for each other. This, it would appear, could result only in the formation of diffusion currents in the two fluids; the two would mix. This result does not follow, so that it would appear that in adhesion Amœba does not play simply the part of a fluid which wets the substratum. As we have seen (p. 165), there is evidence that the adhesion takes place through the mediation of a viscid secretion.

Whatever the nature of the adhesion, we know it exists at one pole of the Amœba and not at the other. Given such chemical differences between the two poles as would produce this difference in adhesion, then locomotion would follow essentially as we find it to occur in *Amœba limax* or *A. verrucosa*. No further properties except those common to fluids would be required.[*] For the determination of the direction and rate of locomotion, the distribution of these chemical differences would be the essential factor.

Caution is necessary, however, in transferring the results of these and other similar experiments to Amœba. The resemblances between the movements of the inorganic drops and those of Amœba show merely

[*] It will be noted that this statement is made for simple locomotion, and does not refer to the formation of pseudopodia.

that the forces acting upon the two have a similar localization and direction, not necessarily that the forces themselves are identical. This caution is emphasized by the fact that drops moving down an inclined plane as a result of the action of gravity have a similar rolling motion. This is well shown in the drops of glycerine or water on the oiled surface, in the experiments just described. Most (though not all) of the details mentioned above, in which the movements of the inorganic drop resemble those of Amœba, may be observed also in drops moving under the influence of gravity. The essential difference between the two sets of experiments is that the action of adhesion, pulling the drop in a certain direction in the one case, is replaced by the action of gravity, pulling in the same direction, in the other case. Corresponding with this difference is the chief difference to be observed in the movements of the drops under the different conditions. In those moving as a result of greater adhesion on one side, the anterior edge is thin and flat, as in Amœba, while in those moving from the action of gravity this is not true.

In Amœba observation shows that we have the one-sided greater adhesion, and the tendency of the lower surface to cling slightly to the substratum, as in the first set of experiments with the inorganic drops. There remains only something corresponding to the surface tension factor, common to both sets of inorganic experiments, to be accounted for in Amœba. Since Amœba acts like a fluid in many respects, there is no *a priori* reason to deny it surface tension, and nothing further is required to produce locomotion. To this there is, however, one objection. This is found in the roughening and wrinkling of the surface at the posterior end as it contracts, and in the similar roughening of a contracting pseudopodium (pp. 160, 168). This is exactly the opposite of what should take place in a fluid contracting as a result of surface tension. In such a case the primary phenomenon is the decrease in surface; the latter should, therefore, remain perfectly smooth, and as small as possible. Of course, surface tension might be replaced in Amœba by a specific property of contractility of some sort, having its seat a little beneath the surface. Locomotion would then take place as in the inorganic drop, and the wrinkling of the outer surface would be accounted for. On the other hand, if we can account for the contractility by a known property of fluids, such as surface tension, our explanation will, of course, be simpler and more probable. By taking into consideration the apparent fact that the outer layer of Amœba is partly fluid, partly solid, I believe that such an explanation, accounting for the roughening as well as the contractility, can be given; this I shall attempt in the next section of this paper (p. 215).

The formation of projections at the anterior edge or side of the inorganic drop, comparable to the formation of pseudopodia in contact with the substratum in Amœba, may also be induced in the oil drops. For this purpose it is necessary to produce a greater adhesion on a small area at one side. A projection is at once sent out here. The movement in sending out such a projection is the same as that to be observed in the formation of a pseudopodium under such circumstances. The projection is thinnest at the tip; its upper surface moves forward and rolls over at the point, while the lower surface is at rest.

FORMATION OF FREE PSEUDOPODIA.

On the other hand, the projection of free pseudopodia into the water cannot be imitated under these conditions. As we have seen, the movement in the formation of a free pseudopodium differs from that in forming a pseudopodium along a surface merely in the fact that in the latter case the contact surface is at rest, while in the free pseudopodium all surfaces move equally, a given point on the surface remaining approximately at the same distance from the tip. In the inorganic drop projections can indeed be formed in which the surface moves in exactly the same manner as in the free pseudopodia of Amœba, but under conditions that are essentially different. If some small object to which the fluid adheres, such as a sliver of wood, is brought into contact with one side of the drop, the fluid flows out over it, and may form thus a long, slender projection. The surface of this projection moves in the same manner as the surface of a pseudopodium in Amœba (p. 153). Thus the surface of a free pseudopodium shows such movements as it would if drawn out by an object to which it adheres at its tip. Since no such object is present, it is clear that the formation of free pseudopodia is not explicable in this manner.

As we have seen above, the locomotion and the formation of pseudopodia in contact with the substratum could, if they stood alone, be considered due to the adherence and spreading out of a fluid on a solid, as maintained by Berthold (1886). But they do not stand alone; we have the additional fact that pseudopodia may be sent out which are not in contact with the substratum. The anterior edge of an Amœba, further, may be pushed out freely into the water as a single pseudopodium. This may frequently be seen in *Amœba limax*. As a reaction to a stimulus the protoplasm may push upward freely as a thick lobe, till the greater part of the substance is transferred upward and the Amœba topples over (p. 184). In the formation of such a lobe, the protoplasm may flow from both ends toward the middle, producing the "heaping up" from which the thick upward projection results (see p. 183). Currents flowing in this manner could not possibly be produced by adherence to the substratum. Again, in an advancing *Amœba angulata*, short triangular pseudopodia are constantly pushed forward, some in contact with the substratum, others not thus in contact, but raised a little above it (Fig. 54). Some, which are at first free, later come in contact. Clearly, *Amœba is able to perform all the activities concerned in locomotion without adherence to a solid*. The adherence is only necessary that there may be a movement from place to place. The case is quite parallel to that of higher organisms, where contact with the substratum is necessary in order that progression may occur, though all the movements concerned in locomotion may be performed without such contact.

We are compelled to conclude, therefore, that in the advancing end of an Amœba or the projecting pseudopodium there is an active movement of the protoplasm, of a sort which has not been physically explained. This involves the general conclusion that no physical explanation is at present possible of the locomotion and projection of pseudopodia in Amœba.

To account for the contraction of the posterior part of the body, on the other hand, possibly the properties common to Amœba with other fluids are sufficient. If surface tension may be considered the cause of the contraction of the posterior part of the body, it is notable that it acts as a constant factor, tending always to decrease the surface as much as possible, not as a variable factor. In other words, there is no indication that local increase or decrease in surface tension (see note, p. 225) plays any part in the production of the movements, as is maintained in the prevailing theories. The part played by surface tension is thus a very subordinate one.

EXPERIMENTAL IMITATION OF MOVEMENTS DUE TO LOCAL CONTRACTIONS OF THE ECTOSARC, AND OF THE ROUGHENING OF THE ECTOSARC IN CONTRACTION.

Besides the sending out of pseudopodia, there are certain other phenomena in the movements of Amœba for which we lack, so far as I am aware, any attempt at a physical explanation. These are the swinging movements, vibrations, and local contractions of pseudopodia, described on pages 177–179, and the roughening of the ectosarc in the contraction of pseudopodia or other parts of the body (p. 168).

These phenomena are in certain details so similar to some that I have observed in inorganic fluids that I believe it worth while to analyze the latter; possibly they give an indication of the direction in which an explanation of the phenomena in Amœba above mentioned may lie.*

---

* In view of the repeated failures of physical explanations in attempting to account for vital phenomena, one does not approach a new attempt of this sort with great confidence. Yet it is desirable that any possibility of this kind should be worked out and submitted to criticism, in order that its truth or lack of truth may be demonstrated.

Swinging or bending movements take place with special frequency while the pseudopodia are withdrawing; in some Amœbæ such movements are an almost constant accompaniment of withdrawal. As it is withdrawn the pseudopodium becomes roughened or warty on its surface, as we have seen, and at the same time bends to one side or the other, or swings back and forth. The impression given is that the outer layer of the pseudopodium has become partially solid. In withdrawing, the solid substance seems to melt gradually away, in a somewhat irregular manner, so as to leave solid masses connected by liquid protoplasm, the projecting solid masses forming the wart-like roughenings of the surface. When this melting away occurs more strongly on one side, the pseudopodium bends at that point, toward the side which has apparently become more fluid.

We have in such a case, if appearances may be trusted, a mass composed partly of solid, partly of fluid. While it is usually admitted that parts of the protoplasm may become solid at times, little attempt has been made to understand protoplasmic movements by studying the physics of such mixtures of solids and fluids.* In certain experiments with inorganic mixtures of this kind, in which movements were produced that resembled those just referred to in Amœba, the writer became convinced of the possible importance of the physics of such mixtures for the understanding of protoplasmic activities.

The experiments in question were concerned with the movements under the action of surface tension of oil drops to which soot had been added for the purpose of rendering the currents visible. When a large quantity of soot was added, the drops became somewhat stiffened, and now showed to a marked degree a mingling of the characteristic properties of fluids and solids.

In one set of experiments clove oil was thus mixed with soot and introduced as drops into a mixture of three parts glycerine and one part 95 per cent alcohol. The drops move about, as a result of local decrease in surface tension, in the same manner as the olive-oil emulsion in Bütschli's celebrated experiments. Much of the soot collects next to the surface of the drop, and becomes massed in certain regions, as a result of the currents, covering these regions with a sort of crust, this crust being formed of separate solid particles. The particles are crowded together as closely as possible, owing to the surface tension of the fluid in which they are floating.† If the particles are not too minute they may project above the surface of the drop, giving it a rough appear-

---
* Some of the experiments of Rhumbler (1898, 1902) deal with such mixtures, though not with a view to an understanding of the movements, but of certain other processes.

† According to the principles set forth by Rhumbler, 1898, p. 332.

ance, similar to that of the withdrawing pseudopodium of Amœba. Such parts of drops or whole drops, so covered, show peculiar properties. The form may be changed by lowering the surface tension locally or by mechanical action from outside, exactly as in a fluid, but there is an inclination to hold a form once received. The tendency to take the spherical form is still somewhat marked, and if the irregular drop is strongly disturbed it frequently slowly becomes spherical. On the other hand, if not strongly disturbed, it may retain almost any form impressed upon it—cylindrical, flattened, irregular, or with long, slender projections. In this power of receiving and retaining an irregular form, yet with a tendency to become spherical, these drops, of course, resemble Amœba. These properties vary with the amount of soot present in the oil; if this is less, the drops slowly return to the spherical shape when deformed; if greater, they retain the irregular shape indefinitely. Such irregular masses nevertheless flow together if brought in contact, will quickly gather together into a close mass if strongly deformed, and in many other ways they show the characteristics of fluids.

The reason for their tendency to retain irregular forms is obvious. The surface is covered with small solid particles that are in contact. The projection of these particles above the surface may cause a roughening of the surface. Any change of form, such as surface tension would produce, causes much friction between these particles. The form taken is, then, a resultant of the action of surface tension and the resistance of these particles to movements.[*] Similar forms are producible by mixing soot with bone oil and studying drops of the mixture in a vessel of glycerine.

For our purpose the phenomena which occur when the soot particles are unequally distributed are of special interest. Consider an elongated projection, as in Fig. 78, A, with the surface entirely covered with closely crowded soot particles except in a certain region, x–y, on one side. In this region x–y surface tension will have free play, tending to draw the points x and y together, while elsewhere the tendency to contraction will be resisted by the friction of the particles. The result is that the points x–y are drawn together, and the projection bends toward that side. Since this bending does not tend to crowd the particles on other parts of the surface closer together they do not resist it.

In the oil drops mixed with soot the bending of projections in the manner described is often to be observed under the appropriate conditions. They should be compared with the bending of pseudopodia as

---

[*] Similar forms of fluids have been produced by Rhumbler in a parallel manner in his imitations of the formation of Difflugia shells (1898, p. 287) and of the shapes of the shells of Foraminifera (1902, p. 265).

described for Amœba (p. 177, and Fig. 62, *a*). The chief experimental difficulty in producing such bending is to arrange the conditions in such a way that there is less soot on one side of a projection than on the other. This occurs somewhat frequently when two irregular drops are allowed to fuse, or when a drop is mechanically deformed with a rod, as described in the next paragraph. One can sometimes bring it about by placing with the capillary pipette a minute drop of oil on one side of a projection, though this method is not as a rule very effective.

Such drops may also show another property in common with Amœba, namely, elasticity of form, or a phenomenon producing similar results. Consider a curved projection, as in Fig. 78, B. It is completely covered with soot particles and retains its form in virtue of their resistance to a change in position. Now, suppose we forcibly straighten out the prosection by pushing it to one side with a rod. By so doing the side *a–b* is lengthened; the soot particles on this side are, therefore, separated, leaving certain areas of free fluid surface. When the projection is released, surface tension can act on these areas, and the projection is drawn back at once to its original form. I have often observed such immediate returns to the original form after bending a projection of one of the oil drops.

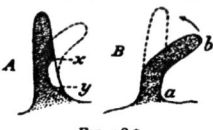

FIG. 78.*

In Amœba we have exactly the conditions most favorable for the production of movements of this sort, and we actually find numerous movements of just this character. It is generally admitted that the outer layer becomes partially solidified; as a pseudopodium is withdrawn the solid portions evidently become liquefied in an irregular way, some of them projecting above the surface and making it rough. If the liquid substance produced shows surface tension, the movements described must follow in the manner set forth above. It seems possible that many of the observed movements are thus produced by local liquefaction, with the intervention of surface tension, in the liquefied area.

In view of the apparently unlimited possibilities of partial solidification and liquefaction in the protoplasmic body, with the resulting varied action of surface tension, shall we not go a step farther and inquire whether there may not be an outlook for an explanation of vibratory movements, such as we find in flagella, along this line? In an elongated structure like a flagellum, a limited liquefaction of one side would result in a bending toward this side. By regular alternation of liquefactions in different regions, a regular vibration could be produced. The chief difficulty in the way of such a theory would seem to lie in

*FIG. 78.—Diagrams illustrating phenomena in mixtures of oil and soot.

the restoration of the original length in a given side after liquefaction and consequent contraction had occurred. This could,.perhaps, be brought about by an elastic rod in the axis of the structure, such as many cilia and flagella are known to possess.

The above is merely a suggestion made tentatively; its justification as a suggestion lies in the following facts: (1) The swinging movement of pseudopodia in Amœba in some cases strikingly resembles movements of the character above set forth in inorganic fluids, and precisely the conditions for such movements are present in Amœba; (2) Swinging movements of pseudopodia seem to grade almost insensibly into the vibratory movements of flagella.

DIRECT OR INDIRECT ACTION OF EXTERNAL AGENTS IN MODIFYING THE MOVEMENTS.

Is the effect of external agents in modifying the movements of Amœba due to the direct physical action of the agent on that part of the fluid substance with which it comes in contact? Or is its action indirect, in that it serves merely as a stimulus to certain internal changes, the latter bringing about the modifications in the behavior? Both views find adherents. The difference between them is fundamental, for they lead to essentially different conceptions as to the nature of behavior in these lower organisms.

In higher animals we know that the movements and changes of movement are not produced in a direct way, but the effect of external agents is to cause internal alterations which result in changes of movement. It is not, therefore, possible to predict the movements of the organism from a knowledge of the direct physical changes produced in its substance by the agent in question. If the theory of direct action is correct for Amœba, we have in these animals a condition of affairs incomparably simpler, for here we can resolve the behavior directly into its physical factors. If, on the other hand, the theory of indirect action is correct, then there appears to be nothing fundamentally different in principle between the behavior of Amœba and that of higher organisms.

How the form and movement of a fluid mass might be determined by the direct action of external agents on its surface may be simply illustrated by certain experiments which I have described elsewhere (Jennings, 1902). A mixture of 2 parts glycerine and 1 part 95 per cent alcohol is placed on a slide and covered with a cover glass supported by glass rods. Into this is introduced with a capillary pipette a drop of clove oil. The clove-oil drop, at first circular in form, soon changes shape, shows internal currents, sends out projections in various directions, moves about from place to place, and may divide into two drops. The alcohol, not being uniformly distributed throughout the glycerine, acts more strongly on some parts of the clove-oil drop than on others, thus lowering the surface tension in the region acted upon. Thereupon the drop sends out a projection on the side affected, and may follow this up by moving in that direction. We may cause the drop to send out projections and move in a certain direction by placing a minute drop of alcohol near one side, or by heating one side; movement takes place toward the side affected. These agents act directly on the surface of the drop, lowering the tension; the movements are a direct consequence of this change. Parallel phenomena may be produced, as Bernstein (1900) has shown, with a drop of quicksilver. If such a drop is placed in 10 per cent nitric acid, in which bichromate of potash has been dissolved, the drop changes form and moves about. If we place near such a drop, in vessel of 10 per cent nitric acid, a crystal of potassium bichromate, the mercury drop moves rapidly over to the crystal. Here, again, the chemical acts directly on the mercury, lowering the surface tension at the region where it comes in contact with it, thus producing the movement.

Certain authors have held that the movements of Amœba are produced in this way by the direct action of external agents decreasing or increasing the surface tension of certain parts of the fluid mass. As an example of the theory of direct action of external agents in controlling the behavior, we may take the view of the reaction of Amœba to chemicals recently given by Rhumbler (1902, p. 384). According to Rhumbler, it is evident that when an Amœba moves toward or away from a certain chemical, the side directed toward the chemical has, in the first case, a lessened surface tension, in the second case an increased surface tension, as compared with the remainder of the body.

The necessary differences of tension on the positive and negative sides may be easily understood from our present standpoint, by holding that a positively acting chemical decreases the surface tension both in the living alveolar system of the cell and especially on the cell surface, upon which it must work most strongly; that a negatively acting chemical, on the other hand, produces an increase of surface tension in the alveoli of the cell, and especially, again, on the cell surface; this increase is the greater, and from a physical standpoint must be the greater, the more the molecules of the chemical affect or modify the tension of the different cell alveoli or different parts of the cell surface (*l. c.*, p. 384).

The explanation of thermotaxis and electrotaxis would be, according to Rhumbler, "exactly the same as for chemotaxis" (*l. c.*, p. 385); thus also as a result of the direct action of external agents. A fuller explanation of the "tropisms" on this basis is given by Rhumbler in an earlier paper (1898, pp. 183, 188).[*]

[*] Rhumbler emphasizes in the paper just cited (1898, p. 184) the importance of "inner disposition" in deciding what effect shall be produced by external agents, but in the tropisms, at least, he considers the action of the external agent to be direct, the inner disposition deciding merely whether the substance of the Amœba is of such a character as to admit of the production of a given definite change in surface tension by the outer agent.

Based similarly on a direct action of external agents is the theory of amœboid movements proposed by Verworn in 1891, and extended in his paper on *Die Bewegung der lebendigen Substanz* (1892) and in the *Allgemeine Physiologie*. In its original form Verworn's theory considers the movements and changes of form to be brought about directly through chemical attraction (Verworn, 1891, p. 105), but in later publications (1892, 1895) the effect of the chemical is considered, as in Rhumbler's theory, to be that of increasing or decreasing the surface tension.

Thus it is evident that it must be the chemical affinity of certain parts of the protoplasm for oxygen that decreases the surface tension in definite regions and thus leads to the formation of pseudopodia. But it will be possible for the same effect to be produced by other substances of the surrounding medium, if they have chemical affinity for certain components of the protoplasm. In the case where the substance acts from one side, this principle must lead to positive chemotropism. (Verworn, 1895, p. 545.)

It is evident that the method of movement of Amœba, as described in this paper, has an immediate bearing on the question of direct or indirect action of external agents. If the action of an external agent is to increase or decrease directly the surface tension, as set forth by Rhumbler and Verworn, this effect must be shown in the characteristic currents which appear in any fluid when the surface tension is thus locally changed. In the case of negative chemotaxis we should have an axial current away from the side affected, with surface currents toward the chemical, as indicated in the figure given by Rhumbler (1898, p. 188). In positive chemotaxis both sets of currents should be the reverse of that just indicated.

In the account of the movements set forth by Bütschli and Rhumbler, the currents agreed with the scheme for direct action above set forth. But this account of the movements was erroneous, as we have seen. The internal currents and the surface currents are forward, away from the region stimulated, in a negative reaction; toward the region stimulated in a positive reaction; the movement is of a rolling character. There is thus no evidence that the action of the stimulus is to cause a change in the surface tension of the parts directly affected; on the contrary, the direction of the currents is quite inconsistent with this view.[*]
We must conclude, then, that the theory of the direct action of external agents in causing or changing the movements of Amœba is negatived by the character of the movements produced; these are not such as would follow from the direct physical action of the agents in question.

---

[*] A description of the forward surface currents in negative chemotaxis is given on p. 143; in the reaction to a mechanical stimulus on p. 185; to the electrical stimulus on p. 192; in a positive food reaction (chemical and mechanical stimuli?) on p. 198.

There is much indirect evidence that points in the same direction, particularly in the fact that the activities of the animal may remain constant while the environment is continually changing. Some of these lines of evidence are summarized by Rhumbler (1898, p. 185). They still leave open the possibility that when the environment does modify the movements its action is direct. But even this is shown to be excluded by the nature of the currents produced, as described above.

The currents as they actually occur in the movements are equally opposed to certain theories of the indirect action of stimuli. Bernstein (1900), Jensen (1901), and others, have expressed the opinion that the effect of stimuli is to change the surface tension, but that this effect is not due to the direct physical action of the agent on the protoplasm, but rather to some change in the internal physiological processes of the cell produced by the agent acting as a stimulus. Jensen (1901, 1902) has developed this view into a detailed theory, according to which stimuli that increase the normal assimilatory processes of the cell lead to a reduction of surface tension, and hence to expansion and movement toward the agent in question, while stimuli that increase the dissimilatory processes have the opposite effect.

The currents in the moving Amœba lend no support to this view. There is no evidence in the movement that the effect of a stimulus is to alter the surface tension in any way. In view of the facts given in the body of this paper as to the nature of the movements, we are forced to give up the idea that the effect of stimuli is to modify the tension* of the surface of the protoplasmic mass, either directly or indirectly. Alterations in the tension of the surface can no longer be considered the prime factor in the behavior of Amœba.

DIRECT OR INDIRECT ACTION IN THE TAKING OF FOOD.

Rhumbler, in his most interesting and suggestive paper (1898), has attempted to give a physical analysis of food-taking and the choice of food in Amœba. According to Rhumbler, the taking of food is due to adhesion between the protoplasm and the food substance, and may be compared with the pulling inward of a splinter of wood by a drop of water, or of a bit of shellac by a drop of chloroform. The selection of food is explained as due to the fact that the protoplasm, as might be expected from physical considerations, tends to adhere to some substances and not to others. Parallel phenomena are shown, in a most ingenious experiment, to be demonstrable for the chloroform drop (*l. c.*, p. 248). It takes in certain substances, while others are refused or thrown out if introduced.

---

*See note, p. 225.

This theory is a most attractive one, and seems *a priori* probable.[*]
It is conceivable that there may be, or may have been, organisms where it is applicable throughout. But an objective study of the behavior of Amœba shows that it gives by no means an adequate explanation of food-taking in this animal. As I have shown in the descriptive portion, in *Amœba proteus* and *A. angulata* the food in most cases is far from adhering to the protoplasm ; on the contrary, it rolls away when the Amœba comes in contact with it, and it is often only as a result of long-continued effort that the animal succeeds in ingesting it. The first act in ingestion consists in sending out pseudopodia on each side of the mass to overcome the mechanical difficulty resulting from the fact that the body does *not* adhere to the protoplasm, but tends to roll away.

Further, a quantity of water is usually, or frequently, taken in with the food, and the latter floats about in a cavity after it is ingested, showing no tendency to adhere to the protoplasm (see Leidy, 1879, numerous figures of food vacuoles, etc., and Le Dantec, 1894). A similar condition of affairs is shown in the account of the feeding of one Amœba on another, given on page 201 of the present paper. Here the prey does not adhere to the protoplasm of its captor, but moves about within the latter and escapes repeatedly.

Thus, in these species, the taking of food and the choice of food cannot be explained by the adherence of the protoplasm to the food substance, for the lack of such adherence is strikingly evident.

Rhumbler's studies of food taking were made chiefly on *Amœba verrucosa*. In this species and its close relatives there is much more tendency for foreign objects to cling to the surface than in the other species. But this adhesiveness applies to other objects as well as to food. It is of special aid, as we have seen, in tracing surface movements (p. 140). Particles of soot and various other indifferent bodies stick to the surface, rendering its movements apparent. Not all such adhering bodies are taken into the interior, so that the ingestion involves an additional reaction, and is not fully accounted for by the adhesion even in these species.

Rhumbler has given an ingenious physical analysis of the rolling up and taking in of filaments of Oscillaria by *Amœba verrucosa*, and has illustrated the process as he conceives it to occur by a very striking experiment (1898, p. 230). A chloroform drop brought in contact with the middle of a filament of shellac rolls the filament together and encloses it. Rhumbler conceives the forces at work in rolling up the filament to be essentially the same in the chloroform drop and in the Amœba. In both cases, according to Rhumbler, the surface tension of

---
[*] See Jennings, 1902, where I adopted this view before having investigated for myself the behavior of Amœba.

the drop pulls on the filament, tending to force it inward from both directions. That part of the filament within the drop becomes softened by the action of the fluid; it, therefore, yields to the thrust from both directions, and bends, permitting more of the filament to be brought into the drop by the action of surface tension. (For full explanation, see the original.)

It is necessary to point out that the explanation of the rolling up of the shellac filament given by Rhumbler is erroneous. The surface tension of the drop, with its inward thrust, has nothing to do with the process, for the filament is rolled up in exactly the same manner when it is completely submerged in a large vessel of chloroform, so that it is not in contact with the surface film at all. The rolling up is evidently due in some way to the strains within the shellac filament, produced when it was drawn out, and to the adhesiveness of its surface when acted upon by the chloroform. The process thus loses all similarity to the rolling up of the alga filament by Amœba. The coil formed is just as small and close, and the filament remains a filament just as long when the experiment is tried in a large vessel of chloroform as when only a drop is used, as in Rhumbler's experiments.

Rhumbler's explanation of the way in which Amœba rolls up the Oscillaria filament may, of course, still be correct, though the physical experiment by which he attempted to illustrate it has nothing to do with the matter. There are certain points in his description of the process as it occurs in Amœba, however, that might easily be interpreted in another manner. Such a bending over of the pseudopodium as is shown in Rhumbler's Fig. 58 (*l. c.*, p. 233) is not called for by the surface-tension theory. Rhumbler holds that this bending of the pseudopodium is passive, and due to the bending of the filament within the body (*l. c.*, p. 233). In view of what we have shown above (pp. 177-179) as to the power of active bending in the pseudopodia, and as to active contractions of parts of the ectosarc in this same species (pp. 179, 180), one might be inclined to believe rather that this bending of the pseudopodium is active and plays an important part in bringing the filament into the body. Rhumbler's figures (Fig. 50) would support this view fully as strongly as his own theory, though this would, of course, not give us a simple physical explanation of the ingestion of the filament.

Altogether, we must conclude that adhesion between the protoplasm and the food substance cannot by any means give us a general explanation of food-taking in Amœba. In some cases the ingestion of food is aided by such adhesion, but in other cases the adhesion is conspicuously absent.[*]

---

[*] For further confirmation of last stated fact, see paper of Le Dantec (1894).

GENERAL CONCLUSION.

Putting all our results together, we must conclude that the movements and reactions of Amœba have as yet by no means been resolved into their physical components. Amœba is a drop of fluid which moves in its usual locomotion in much the same way as inorganic drops move under the influence of similarly directed forces. But what these forces are is by no means clear. When we take into consideration the currents as they actually exist, local decrease in surface tension breaks down completely as an explanation for the locomotion and other movements. The locomotion taken by itself might be explained as due to the adhesion of the fluid protoplasm to solids, taken in connection with the surface tension of the fluid, but this explanation fails when we consider the formation of free pseudopodia, and discover that all the processes concerned in locomotion can take place without adhesion to the substratum.

For the reactions to stimuli we find a parallel condition of affairs. The currents in the protoplasm in the positive and negative reactions are not similar to those produced in the attraction or repulsion of drops of fluid by the direct action of external agents. Therefore we cannot consider these reactions as due to the increase or decrease of surface tension* produced by the direct (or even indirect) action of the external agents. The taking and choice of food cannot be physically explained in any general way by the physical adherence of the protoplasm as a substance to the food as a substance, for food is taken in many cases (usually, in some species) where it is demonstrable that no such adherence exists.

While we must agree that Amœba, as a drop of fluid, is a marvellously simple organism, we are compelled, I believe, to hold that it has many traits which are comparable to the "reflexes" or "habits" of higher organisms.† We may, perhaps, have faith that such traits are

---

*It should be pointed out that this and other statements concerning surface tension in Amœba apply to the tension of the actual body surface, comparing Amœba thus to a drop of simple fluid. This is the basis on which rest the prevailing theories that would explain the movements of Amœba by surface tension. It is these theories which I desired to test. There remains untouched, of course, the possibility that the movements of all sorts of protoplasmic masses may be explained by changes in the surface tension of the meshes of Bütschli's honeycomb structure, in the manner indicated by Bütschli (1892, p. 208). But this is at present merely a hypothesis, not worked out and not controllable by observation. To attempt to maintain it for Amœba would be to relegate the movements of this animal to the same obscure category as the movements of cilia and of muscles, possibly a correct proceeding, but removing the matter at present from the field of experimental observation.

† See the next division of this paper, where this point is developed.

finally resolvable into the action of chemical and physical laws, but we must admit that we have not arrived at this goal even for the simpler activities of Amœba.

THE BEHAVIOR OF AMŒBA FROM THE STANDPOINT OF THE COMPARATIVE STUDY OF ANIMAL BEHAVIOR.

HABITS IN AMŒBA.

Although in general Amœba has the rolling movement of a drop of fluid, yet this statement by no means brings out all the characteristics of the movement in any given species of Amœba. Different kinds of Amœbæ move differently, and the differences are in many cases not such as can be accounted for by differences in the state of aggregation of the body substance. Some Amœbæ, as is well known, form many pseudopodia, others few or none. Different Amœbæ have different characteristic forms in locomotion. But more striking than these generally recognized peculiarities are certain others of a more special character. A creeping *Amœba angulata*, as we have seen above, frequently pushes upward and forward at the anterior end a short, acute pseudopodium, which waves slightly from side to side like an antenna (p. 177 and Fig. 62, c). This peculiar habit is much more pronounced in *Amœba velata* Parona, according to Penard (1902). In this animal the free anterior pseudopodium may extend for a length greater than the diameter of the body; Penard compares it directly to a tentacle. Some other species of Amœba never send forward such an antenna-like pseudopodium. The great work of Penard (*l. c.*) contains innumerable instances of such peculiarities of form, movement, and function among the different species of Amœba and other Rhizopods; some of them are collected in that author's interesting section on the pseudopodia (*l. c.*, pp. 625–629). It is not necessary to take these up in detail here. The point of interest is that different sorts of Amœbæ have different customary methods of action, such as are commonly spoken of as " habits "* in higher animals, and that these " habits " are no more easily explicable on direct physical grounds in Amœba than in higher animals. Let anyone attempt, for example, to explain from the physics of viscous fluids why *Amœba velata* or *A. angulata* push out an antenna-like pseudopodium at the anterior end and wave it from side to side, while *Amœba proteus* and *A. limax* do not.

---

*The word *habit* is, of course, not used here of a method of action acquired during the life of the individual, but merely of a fixed method of behavior. At all events, it is difficult to distinguish between these two things where individual organisms, as in Amœba, have lived as long as the race.

CLASSES OF STIMULI TO WHICH AMŒBA REACTS.

The simple naked mass of protoplasm reacts to all classes of stimuli to which higher animals react (if we consider the auditory stimulus merely a special case of the mechanical stimulus). Mechanical stimuli, chemical stimuli, temperature differences, light, and electricity—all control the direction of movement, as they do in higher animals.

TYPES OF REACTION.

Amœba has two chief types of reaction, by one or the other of which it responds to most stimuli. These we may call the positive and the negative reactions. As a third type we must distinguish the food reaction, which cannot be brought completely under either of the two chief types of reaction above mentioned.

(1) The positive reaction consists in pushing out the body substance toward the source of stimulus and rolling in that direction.

(2) The negative reaction consists in withdrawal of body substance and rolling in some other direction—not necessarily in the opposite direction.

(3) The food reaction is not sharply definable. Its most characteristic features consist in the hollowing out of the anterior end and in the pushing out of pseudopodia at each side of and over the food body. It involves also the positive reaction above characterized.

RELATION OF THE DIFFERENT REACTIONS TO DIFFERENT STIMULI; ADAPTATION IN THE BEHAVIOR OF AMŒBA.

(*a*) The positive reaction is known to be produced by weak mechanical stimuli; it is probably produced also by weak chemical stimuli (in the reactions to food, pp. 193-202). The positive reaction to weak mechanical stimuli serves the purpose of bringing the floating animal to a surface on which it can creep. The positive reaction to food substances (mechanical and chemical stimuli), of course, serves to obtain food. The positive reaction is thus, as a rule, performed under such circumstances as to be beneficial to the organism; *i. e.*, it is directly adaptive.

(*b*) The negative reaction is produced by powerful stimuli of all sorts. Such powerful stimuli are, as a rule, injurious, and the negative reaction tends to remove the Amœba from their action; it is, therefore, directly adaptive. This is true of the negative reaction to light as well as to other stimuli, for light is known to interfere with the activities of Amœba. The reaction to the electric current is of exactly the character that would be produced by a strong stimulus on the anode side, but owing to the peculiarity of the current the reaction does not assist the Amœba to escape. The reaction to the electric current can not then be considered adaptive; this stimulus forms, of course, no part of the normal environment of an Amœba.

228   THE BEHAVIOR OF LOWER ORGANISMS.

The method by which, through the negative reaction, Amœba avoids an injurious agent is of interest. The animal does not go directly away from the injurious agent, as by moving toward the side opposite that stimulated. It moves in any direction except toward the region stimulated. There is no system of conduction such that strong stimulation in a given spot involves movement directed toward the opposite side. If the movement in the new direction induces a new stimulation, the direction is again changed. This may be continued until the original direction of movements is squarely reversed. The method is akin to that of trial and error in higher organisms (see Morgan, 1894, pp. 241-242).

(c) The food reaction is directly adaptive in that it procures food. As a rule this reaction occurs only when the source of stimulation is fitted to serve as food. Empty diatom shells, sand grains, débris, etc., are, as a rule, not taken into the body, as many observers have pointed out. Sometimes material is taken into the body that is not useful, as is described by Rhumbler (1898, p. 236). In such cases there is no evidence of a food reaction in the sense characterized above; the material is ingested accidentally, as it were, through its adherence to the protoplasm. The food reaction, as a definite form of behavior, is always adaptive so far as known.

(d) Some of the habits of Amœba, characterized above, are clearly adaptive. The use of the antenna-like pseudopodium sent out by *Amœba velata* and *A. angulata* is evident. Penard describes in detail how *Amœba velata* uses it in passing from one substratum to another.

The habit which some Amœbæ have, when suspended freely in the water, of sending out pseudopodia in all directions (p. 181) is, of course, useful in that it increases the chances of coming in contact with some solid object, without which the Amœba cannot move from place to place.

REFLEXES AND " AUTOMATIC ACTIONS " IN AMŒBA.

In the behavior of Amœba we can distinguish factors directly comparable to the reflexes and "automatic activities" of higher organisms. The responses of Amœba to stimuli have the nature of reflexes in the fact that they are not direct effects of the physical action of the stimulus (see p. 219), but are determined by the internal conditions of the organism. They may be called reflexes, unless we propose, as certain writers do, to restrict the term reflex to processes involving differentiated nerves. The precise designation is unimportant; the essential point is that the responses agree with the reflexes of higher animals in being indirect.

Ziehen, in his *Leitfaden der physiologischen Psychologie* (sixth edition, p. 10), defines as automatic acts " motor reactions, which do not, like the reflexes, follow unchangeably upon a definite stimulus, but which are modified in their course by new, intercurrent stimuli." In this sense Amœba, of course, shows automatic behavior. Its responses are by no means unchangeably fixed; on the contrary, its behavior is often modified by the slightest change in the stimulus to which it is reacting. For examples of this see the chase of one Amœba by another (p. 200), the following of a rolling ball of food (p. 196), the account of the driving of Amœba (p. 185), and the description of the method by which Amœba avoids an obstacle (p. 186).

Whether these actions agree with the accepted idea of an automatic action in being unconscious we have, of course, no means of knowing.

VARIABILITY AND MODIFIABILITY OF REACTIONS.

There is little that can be said on this point. Verworn (1890, a, p. 271) says that when an electric current is passed through a preparation containing many Amœbæ, some respond strongly, while others do not; thus different individuals vary in their responsiveness. Further, a given individual may become accustomed to the current, at first responding to it, later not responding. Doubtless such phenomena of acclimation are common in the reactions to all sorts of stimuli.

Rhumbler (1898, p. 203) shows that when Amœbæ are engaged in taking Oscillaria filaments as food, light thrown upon them modifies them physiologically in such a way that they eject the food. The nature of the reaction is thus shown to depend partly on the physiological condition of the animal.

There is no direct experimental evidence as yet, so far as I am aware, that Amœba shows memory.[*] Experimental evidence as to whether the reactions of a given Amœba to a given stimulus are modified by previous stimuli received is very difficult to obtain, principally because it is practically impossible to make succeeding stimuli alike, so that one cannot tell whether a difference in the reaction is due to a difference in the present stimulus or not. Possibly there is a faint indication of something akin to memory shown in the facts described on page 201. Here a smaller Amœba which had been ingested as prey escaped from the posterior end of the captor; the latter thereupon reversed its movements, came up with the escaping prey, and again ingested it. In the interval between the complete separation of the prey from its captor and its recapture, the behavior of the captor would seem to have been determined by some trace left within it by the former possession of the

[*] The word memory is, of course, used here of the objective phenomenon that in many animals present behavior is modified in accordance with past stimuli received, or past reactions given. Of possible subjective accompaniments of this objective phenomenon we, of course, know nothing directly so far as the lower organisms are concerned.

prey. But, possibly, this trace was merely of a gross physical character, acting as a direct stimulus to produce the observed behavior. If this is true, the behavior shows no indication of memory.

SUMMARY.

OBSERVATIONS.

THE USUAL MOVEMENTS.

(1) Locomotion in Amœba is a process that may be compared with rolling, the upper and lower surfaces continually interchanging positions. This is shown by observation of the movements of particles attached to the outer surface or embedded in the ectosarc of the animal. Such attached particles move forward on the upper surface and over the anterior edge, remain quiet on the under surface till the body of the Amœba has passed, then pass upward at the posterior end and forward on the upper surface again. Single particles may thus be observed to make many complete revolutions. (See p. 170, Fig. 58, and Figs. 38, 39, 40, 41.) *

(2) Thus the upper surface moves forward in the same direction as the internal currents, while the lower surface is at rest. There is characteristically no backward current anywhere in Amœba, though at times some of the endoplasmic particles, spreading out laterally at the anterior end, may move a slight distance backward at the sides. This is rare (see p. 134). The forward current on the upper surface is not confined to a thin layer, but extends inward to the endosarc; the endosarcal and surface currents are one (p. 142).

(3) In the formation of pseudopodia that are in contact with the substratum the movement of protoplasm is identical with that at the anterior end of the Amœba. The upper surface and internal contents flow toward the tip, while the surface in contact with the substratum is quiet. Particles adhering to the upper surface are carried out to the tip and rolled under to the lower surface, where they remain quiet (p. 152).

(4) In the formation of pseudopodia projecting freely into the water, the movements of substance are the same as in pseudopodia that are in contact, save that there is no part of the surface at rest. The whole

---

* Of anyone who is inclined to reject these results on the basis of previous observations, or of their supposed incompatibility with other known facts, let me make the following request: Before taking ground against the results, procure some specimens of *Amœba verrucosa* or one of its relatives. This is usually easily done. Then mix thoroughly with the water containing them some fine soot, and observe carefully the movements of the animals. The particles attach themselves to the outside, and the movements of the surface are then observable with the greatest ease. It is such a simple matter to determine certain of the chief points for one's self in this manner that it would be regrettable for controversy to arise through neglect of the needed observations.

surface thus moves outward, and new parts of the surface of the body continually pass on to the pseudopodium. An object adhering to the surface of the pseudopodium remains at approximately the same distance from the tip, both when the pseudopodium is short and when it has become very long (pp. 153–156, and Figs. 47–49).

(5) In the withdrawal of a free pseudopodium (*a*) a process occurs that is the reverse of that described in (4), the surface of the pseudopodium passing from its base on to the surface of the body (p. 156, and Fig. 49); (*b*) the surface of the pseudopodium becomes wrinkled and shrunken; (*c*) the endosarc flows back into the body.

'(6) Any part of the protoplasm may be excluded temporarily from the forward currents. In many Amœbæ there is usually a region at the posterior end which is thus temporarily excluded (the posterior appendage, tail). In such cases the lower surface of the Amœba passes upward on each side of this appendage to become part of the upper surface, then passes forward (p. 169, and Fig. 57). The substance of the posterior appendage is itself gradually drawn into the forward current.

(7) The anterior portion of the advancing Amœba is attached to the substratum, while the posterior portion is not (p. 165). There is a viscid secretion produced on the outer surface of the Amœba, to which the attachment may be due.

(8) The attached anterior portion of the body is spread out and usually very thin. The unattached posterior portion becomes rounded and thick, and is contracting, so that there is a slight forward movement on the lower surface, as well as on the upper surface, in this part (p. 166).

(9) All the activities concerned in locomotion can be performed when the animal is not attached to the substratum. (But for progression such attachment is necessary, p. 215.)

(10) The locomotion of Amœba is similar even in details to the movements of a drop of inorganic fluid which adheres strongly to the substratum at one edge and spreads out upon it here, while the other edge is free (pp. 209–214). It is similar in most respects (except in the thinness of the anterior edge) to the movements under the influence of gravity of a drop of fluid along an inclined surface to which it adheres but slightly.

(11) The currents in a moving Amœba are not similar to those of a drop of inorganic fluid that is moving or elongating as a result of a local increase or decrease in surface tension. The surface currents away from the region of least tension and in the opposite direction to the axial currents that are characteristic of such a drop are lacking in Amœba. Here surface and axial currents have the same direction (p. 205).

(12) Similarly, the movements of material in a forming pseudopodium are not like those in a projection which is produced in a drop of inorganic fluid as a result of a local decrease in surface tension. The surface currents away from the tip in the inorganic projection are lacking in Amœba, the surface here moving in the same direction as the tip (p. 206).

(13) The body of Amœba shows under some conditions elasticity of form, such as is characteristic of solids (pp. 175–177).

(14) Besides the movements directly concerned in the usual locomotion, limited local contractions may occur, resulting in swinging or vibratory motions of the pseudopodia (p. 177), or in sudden, jerky movements of the body as a whole (p. 179), or in the constricting off of parts of the body (p. 180).

(15) The roughness of surface in a retracting pseudopodium, the retention of irregular forms by Amœba, and the movements mentioned in the foregoing paragraph, are similar to certain phenomena to be observed in mixtures of solids and fluids, as a result of the interaction of surface tension and the friction of the solid particles (p. 215).

REACTIONS TO STIMULI.

(16) The following reactions are described: Positive and negative reactions to mechanical stimuli (pp. 181–187); negative reactions to chemical stimuli (pp. 187–190); negative reaction to heat (pp. 190–191); reaction to the constant electric current (pp. 191–192); complex reactions connected with food-taking (pp. 193–202); reactions to injuries (pp. 202–204).

(17) In most reactions modifying the direction of motion a new advancing wave of protoplasm is sent out from some part of that portion of the body which is already attached to the substratum. The internal and surface currents then flow in that direction, thus changing the course of the animal. Thus, when the animal changes its course in a reaction, the surface currents change their course in a corresponding way, as shown by the movements of particles on the surface (pp. 143, 185, 192).

(18) Sometimes the course is squarely reversed as a result of a stimulus. In this case the original anterior region becomes detached from the bottom, while a new pseudopodium projects freely into the water from the former posterior (unattached) region, settles down, and becomes attached; the internal and surface currents then follow it. This process requires usually a considerable interval of time (pp. 183, 184).

(19.) Both surface currents and internal currents are toward the source of stimulation in a positive reaction, away from the source of stimulation in a negative reaction.

(20) The currents in the positive and negative reactions are not similar to the currents in a drop of inorganic fluid moving toward or away from an agent which causes a local decrease or increase in the surface tension. In Amœba the currents on the surface and in the interior are congruent; in the inorganic fluid they are opposed.

(21) In the taking of food the protoplasm and the food body in many cases do not tend to adhere, so that the Amœba is compelled to overcome considerable mechanical difficulty before the food can be inclosed. Frequently the food body rolls away from the animal as soon as it is touched (pp. 193, 196). The difficulty is overcome by sending out pseudopodia on each side of the body and inclosing it, together with a certain amount of water. In *Amœba verrucosa* and its relatives food-taking is aided by the tendency of foreign bodies to adhere to the body surface. Amœbæ frequently prey upon each other, and this often gives rise to a long and complex train of reactions (pp. 198–202, and Fig. 76).

CONCLUSIONS.

(22) The chief factors in locomotion seem to be as follows: (1) At the anterior edge of the Amœba a wave of protoplasm pushes out, rolls over, and becomes attached to the substratum; (2) This pulls on the upper surface of the Amœba, causing it to move forward; (3) The hinder portion of the Amœba becomes released from the substratum, and contracts slowly; (4) As a result of the strong pull from in front and the slight contraction from behind the posterior end moves forward; (5) The internal substance must flow forward as a result of the pull on the upper surface, the movement forward of the posterior end, and the pressure due to the pulling from in front and the contraction behind. The movement of the internal fluid is comparable to that in a sac or bladder half filled with water and rolled along a surface (pp. 146, 149, 171).

(23) There is no continuous transformation of endosarc into ectosarc at the anterior end, and of ectosarc into endosarc behind this (Rhumbler's ento-ectoplasm process), as a necessary feature of locomotion, since the ectosarc of the upper surface rolls over to the under surface at the anterior end (p. 148). Nevertheless, ectosarc and endosarc are mutually interconvertible when need arises for the change of one into the other.

(24) It results from paragraphs (1), (2), (11), above, that the locomotion of Amœba cannot, with fidelity to the results of the physical experiments, be accounted for by a decrease in surface tension at the anterior end.

(25) From (3), (4), (12), above, we must conclude that the sending out of pseudopodia cannot, without violence to the results of the physical experiments, be accounted for as due to a local decrease in surface tension at the point of the pseudopodium.

(26) From (1) and (10) above it results that the simple locomotion on a substratum could, taken by itself, be accounted for on Berthold's theory that the movement is due to the spreading out of a fluid on a solid. But this theory fails when we take into account the formation of free pseudopodia (p. 214), and the fact that all the processes concerned in locomotion can be performed without adherence to a solid (paragraph (9) above).

(27) From (3), (4), (9), (11), (12), (24), (25), (26), we must conclude that the formation of pseudopodia and the sending out of waves of protoplasm at the anterior end of a moving Amœba are due to a local activity of the protoplasm for which no physical explanation has been given. Since these are the essential features in locomotion, we must conclude that locomotion in Amœba has not been physically explained.

(28) From (17), (19), (20), it follows that we cannot, with fidelity to the results of physical experimentation, hold that the effects of stimuli in modifying the movements of Amœba are due to their direct (or even indirect) action in changing the surface tension of the parts affected.

(29) From (21) we must conclude that adherence between the protoplasm and the food substance does not furnish an adequate explanation of food-taking and the choice of food in Amœba.

(30) From (24), (25), (27), (28), we must conclude that changes in the surface tension of the body are not the primary factors in the movements and reactions of Amœba. (See note, p. 225).

(31) All the results taken together lead to the conclusion that neither the usual movements nor the reactions of Amœba have as yet been resolved into known physical factors. There is the same unbridged gap between the physical effect of the stimulus and the reaction of the organism that we find in higher animals.

(32) In the behavior of Amœba we may distinguish factors comparable to the habits, reflexes, and automatic activities (Ziehen) of higher organisms (pp. 228-229). Its reactions as a rule are adaptive (pp. 227-228).

SEVENTH PAPER.

THE METHOD OF TRIAL AND ERROR IN THE BEHAVIOR OF LOWER ORGANISMS.

THE METHOD OF TRIAL AND ERROR IN THE BEHAVIOR OF LOWER ORGANISMS.

A certain type of behavior in higher animals has been characterized by Lloyd Morgan as the method of trial and error. The nature of such behavior is well brought to mind by an example from Morgan (1894, p. 257). His dog was required to bring a hooked walking stick through a narrow gap in a fence. The dog did not pause to consider that the stick would pass through the narrow opening only if taken by one end and pulled lengthwise. On the contrary, he simply seized the stick in the way that happened to be most convenient, near its middle, and tried to carry it through the gap in the fence in that manner. Of course, the stick would not pass, and after some effort the dog was forced to drop it. Then he seized it again at random, and made a new effort. Again the stick was stopped by the fence; again the dog dropped it, took a new hold, and tried again. After several repetitions of this performance, the dog seized the stick by the hooked end. This time it passed through the gap in the fence easily.

The dog had "tried" all possible methods of pulling the stick through the fence. Most of the attempts showed themselves to be "error." Then the dog tried again, till he finally succeeded. Thus he worked by the method of trial and error.

This method of reaction has been found by Lloyd Morgan, Thorndike (1898), and others, to play a large part in the development of intelligence in higher animals. Intelligent action arises as follows: The animal works by the method of trial and error till it has come upon the proper method of performing an action. Thereafter it begins with the proper way, not performing the trials anew each time. Thus intelligent action has its basis in the method of "trial and error," but does not abide indefinitely in that method.

Behavior having the essential features of the method of "trial and error" is widespread among the lower and lowest organisms, though it does not pass in them so immediately to intelligent action. But like the dog bringing the stick through the fence the first time, they try all ways, till one shows itself practicable.

This is the general plan of behavior among the lowest organisms under the action of the stimuli which pour upon them from the surroundings. On receiving a stimulus that induces a motor reaction, they try going ahead in various directions. When the direction followed leads to a new stimulus, they try another, till one is found which does not lead to effective stimulation.

This method of trial and error is especially well developed in free-swimming single-cell organisms—the flagellate and ciliate infusoria—and in higher animals living under similar conditions, as in the Rotifera. In these creatures the structure and the method of locomotion and reaction are such as to seem cunningly devised for permitting behavior on the plan of trial and error in the simplest and yet most effective way.

These organisms, as they swim through the water, typically revolve on the long axis, and at the same time swerve toward one side, which is structurally marked. This side we will call X. Thus the path becomes a spiral. The organism is, therefore, even in its usual course, successively directed toward many different points in space. It has opportunity to try successively many directions though still progressing along a definite line which forms the axis of the spiral (see Fig. 79). At the same time the motion of the cilia by which it swims is pulling toward the head or mouth a little of the water from a slight distance in advance (Fig. 79). The organism is, as it were, continually taking "samples" of the water in front of it. This is easily seen when a cloud of India ink is added to the water containing many such organisms.

At times the sample of water thus obtained is of such a nature as to act as a stimulus for a motor reaction. It is hotter or colder than usual, or contains some strong chemical in solution, perhaps. Thereupon the organism reacts in a very definite way. At first it usually stops or swims backward a short distance, then it swings its anterior end *farther than usual toward the same side X to which it is already swerving.* Thus its path is changed.

FIG. 79.*

After this it begins to swim forward again. The amount of backing and of swerving toward the side X is greater when the stimulus is more intense.

* FIG. 79.—Spiral path in the ordinary swimming of Paramecium, showing how the anterior end is pointed successively in different directions, and how a sample of water is drawn to the anterior end and mouth from each of these directions.

This method of reaction seems very set and simple when considered by itself. It is almost like that of a muscle which reacts by the same contraction to all effective stimuli. The behavior of these animals seems, then, of the very simplest character. To practically all strong stimuli they react in a single definite way.

But if we look closely at this simple method of reacting, we find it, after all, marvelously effective. The organism, as we have seen, is

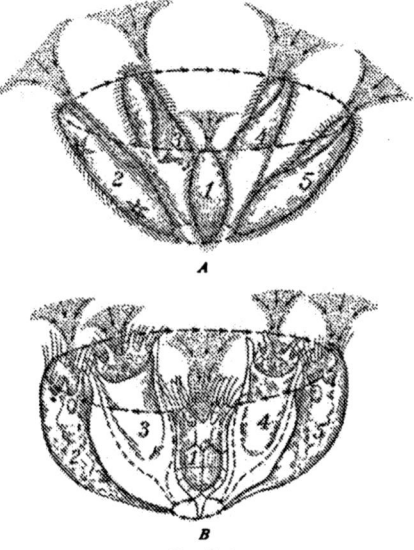

FIG. 80.*

revolving on its long axis. When, as a consequence of stimulation, it swings its anterior end toward the side X, this movement is combined

* FIG. 80 —Diagrams of the movements in a reaction to a stimulus in an infusorian, Paramecium (A), and in a rotifer, Anuræa (B). The anterior end swings about in a circle (turning continually toward the aboral or dorsal side). It thus tries many different directions, at the same time receiving samples of water from each of these directions. 1, 2, 3, 4, 5, the successive positions taken, with the currents of water at the anterior ends. If the stimulus ceases the organism may stop in any of these positions, and swim forward in the direction so indicated. (The backward swimming, which precedes or accompanies the turning, is not represented.)

with the revolution on the long axis. As a consequence, the anterior end is swung about in a wide circle: the organism tries successively many widely differing directions (Fig. 80). From each of these directions, as we have seen, a sample of water is brought to the sensitive anterior end or mouth. Thus the reaction in itself consists in trying the water in many different directions. As long as the water coming from these various directions evinces the qualities which caused the reaction—the greater heat or cold or the chemical—the reaction, with its swinging to one side, continues. When a direction is reached from which the water no longer shows these qualities, there is no further cause for reaction; the strong swerving toward the side X ceases, and the organism swims forward in the direction toward which it is now pointed. It has thus avoided the region where the conditions were such as to produce stimulation.

While the account just given shows the essential features of the reaction, the actual series of events appears in many cases more complicated, though there is nothing differing in principle from what was just set forth. The apparently greater complication arises from the repetition of that feature of the reaction which consists in swimming backward, and in the cessation of the reaction at intervals, with an attempt to swim forward. After the organism has swung the anterior end to one side, if it still receives the stimulus it may begin the reaction anew; that is, it may swim backward a distance, and again begin turning toward the side X. This may be repeated several times. Each time it is repeated the organism swings its anterior end through a new series of positions, thus increasing the chances of finding one in which there is no farther stimulation. Again, the organism, after reacting in the way described in the last paragraph, may begin to swim forward, only to find that it receives the stimulus again; it then repeats the whole reaction, thus supplying itself with a completely new set of directions and of samples of water from those directions. In some cases the reaction is thus repeated many times before any direction is found toward which the organism can swim without receiving stimulation.

This is the method of behavior which the present author has been describing in detail in many organisms in his series of ten Studies on Reactions to Stimuli in Unicellular Organisms,[*] and in the foregoing Contributions to the Study of the Behavior of the Lower Organisms. Not until recently, it must be confessed, has the real significance of this type of behavior been fully perceived. The results seemed to a large degree negative; the reaction method clearly did not agree with the prevailing tropism theory, nor with any other of the commonly

---

[*] Journ. of Physiol., 1897, vol. 21; Amer. Journ. of Physiol., vols. 2 to 8, 1899 to 1902; Amer. Naturalist, vol. 33, 1899; Biol. Bull., vol. 3, 1902.

held theories as to the reactions of lower organisms. Just what the organism did was, indeed, fairly clear, but the plan of it all, the general relations involved in all the details, was *not* clear. This was partly due, perhaps, to overemphasis of certain phases of the reaction and to a tendency to consider other features unimportant. The behavior under stimuli is a unit; each factor must be considered in connection with all the others; then the general method running through it all becomes strikingly evident.

Let us now return to the organisms. Sometimes stimuli are received of such a nature that their distribution is not affected by the currents produced by the cilia; in other words, they cannot be sampled in the currents of water brought to the anterior end or mouth, as shown in Figs. 79 and 80. This is true, for example, of stimulation by light, and of stimulation by contact with solid objects. Under such stimulation the behavior is nevertheless still by the method of trial and error. Let us consider first the reaction to a mechanical stimulus.

When the organism comes in contact with a mechanical obstacle the reaction is exactly the same as that already described. It swims backward, swings toward the side X, and this, with the revolution on the long axis, points the anterior end successively in many different directions. The organism then follows one of these directions. If this leads against the obstacle, the reaction is repeated, till finally a direction is found in which the obstacle is avoided.

In the reaction to light, as it occurs in Stentor or Euglena, experiment shows that changes in the intensity of illumination at the sensitive anterior end are the agents causing reaction (see the second of these contributions). The reaction produced is that already described; by turning toward the side X and revolving on its long axis, the organism tries many directions.

When a negative organism, such as Stentor, comes in its swimming to an area that is more brightly illuminated, or when a positive organism, such as Euglena, comes to an area that is less brightly illuminated, the change in intensity acts as a stimulus. The organism responds in the way already described; it backs away, then tries many different directions by swinging its anterior end about in a circle. It then starts forward in one of these directions. If this does not lead into the area causing stimulation, well and good; if it does, the organism repeats the reaction, trying a new set of directions, till it finds one that does not carry it to the area causing stimulation.

When light coming from a certain direction falls upon one side of a swimming infusorian, the spiral path followed, of course, causes the anterior end to be pointed successively in different directions. As a result, the illumination of the anterior end is repeatedly changed, since in some directions it is pointed more nearly toward the light, in others more away from the light, so as to be partly shaded by the rest of the body. These changes in illumination cause the reaction; the organism tries pointing in various directions. When it comes into such a position that the anterior end is no longer subjected to changes in intensity of illumination, it continues to swim forward in that direction. Such a position is found only when the axis of the spiral path is in the direction of the light rays. In an organism which reacts when the intensity of illumination is decreased, such a position is stable only when the anterior end is directed toward the source of light; in an organism which reacts when the intensity of illumination is increased, only when the anterior end is directed away from the source of light (details in second of these contributions). Thus the organism tries various directions till one is found which does not subject it to changes in intensity of illumination at the anterior end; in this direction it swims forward.

The reaction which produces orientation to light can be stated more simply, but less completely and accurately, as follows: When light coming from a certain direction falls upon the sensitive anterior end of a negative organism, such as Stentor, this causes the reaction above described. The animal, after backing, tries a new set of directions, by whirling its anterior end about in a circle. It continues or repeats this until a direction is found in which the light no longer falls on the sensitive anterior end. It is then oriented with anterior end away from the source of light. In the positive organisms, such as Euglena, the method of reaction is the same, save that it is the shading of the anterior end that causes the reaction. When the anterior end is shaded the organism reacts in the usual way. It tries successively many different directions, by whirling its anterior end about in a wide circle; when the anterior end becomes pointed toward the source of light, the organism continues forward in that direction.

In those infusoria which creep along the bottom, as Stylonychia or Oxytricha, the reaction method is of a slightly simpler character, though identical in principle. These animals when creeping do not rotate on the long axis. When stimulated in any of the ways described, they dart back, then turn to their right. They thus keep in contact with the bottom, and may turn through any number of degrees up to 360 or more (Fig. 81). The reaction places them thus successively in every position with reference to the source of stimulus that is possible so long as they remain on the bottom, and in each position the adoral cilia are bringing samples of water to the anterior end and mouth, as in Fig. 81. When a position is reached where the stimulus no longer acts, the reaction ceases, and the animal moves forward in that direction. The reaction is sometimes repeated several times before the definitive position is attained.

In no other group of organisms does the method of trial and error so completely dominate behavior, perhaps, as in the infusoria. In this group the entire organization seems based on this method. But reactions on this plan are found abundantly elsewhere. In Amœba the present author has shown that many of the reactions are of this character. (See the preceding paper on the movements and reactions of Amœba.) When stimulated mechanically or by a chemical, the Amœba does not move directly away from the source of stimulus, but merely in *some other direction* than that toward the side stimulated. If this leads to a new stimulus, the animal tries another direction. By continued stimuli Amœba may be driven in a definite direction. The conditions necessary for this are that movement in any other direction shall lead to stimulation.

Reaction on the plan of trial and error is, perhaps, best seen in Amœba in the method by which a specimen suspended in the water finds and attaches itself to a solid object. The suspended Amœba sends out pseudopodia in all directions. If the tip of one of these pseudopodia comes in contact with a solid object it becomes attached; the protoplasm begins to flow in that direction, and all the other pseudopodia are withdrawn. The Amœba then passes to the solid and creeps over its surface. Thus the Amœba has tried sending out pseudopodia in all directions; that which has been successful in finding a solid determines the direction of movement.

FIG. 81.[*]

In bacteria the reactions to light, to chemicals, and to mechanical stimuli are essentially like those of the infusoria. The details as to the direction of turning, etc., are not known, owing to the minuteness of these organisms. But the essential point is that when the bacteria are stimulated effectively they change the direction of movement. Such change is repeated until the organisms are brought into a position where there is no effective stimulation. The behavior is clearly that of trial and error.

[*] FIG. 81.—Different positions occupied in the usual reaction to stimuli in Oxytricha. The animal swings its anterior end in a circle, occupying successively positions 1, 2, 3, 4, 5, 6, and receiving a sample of water from each direction in which the anterior end is pointed. When the stimulus ceases the animal may swim forward in any of these directions. (The backward swimming which precedes or accompanies the turning toward the right side is not represented.)

In the Metazoa behavior is usually not that of trial and error in so elementary a form as is found in the organisms thus far considered. The higher animals, with the development of a nervous system and other bodily differentiations, have usually acquired the power of reacting more precisely with reference to the localization of the source of stimulus. They more often, therefore, turn directly toward or away from the source of stimulus, a preliminary trial of different directions being unnecessary. But with the acquirement of many reaction possibilities, the field for the operation of the method of trial and error is greatly broadened. This could be amply illustrated from the behavior of certain of the lower Metazoa. I hope to develop this point in detail at some later time. Since at present we are interested chiefly in the lowest organism, I shall mention here only a few cases in the Metazoa.

In the Rotifera behavior is, under many conditions, precisely similar to that which I have described above for the infusoria (details in the third of these contributions); that is, the behavior is an example of the method of trial and error in a very pure form.

In Hydra we find the method of trial and error in a number of features of the behavior. Since a general paper on the behavior of Hydra is in preparation,[*] I will mention only one or two points that have already been described. If we observe a living, unstimulated green Hydra, we find that it does not remain at rest. If the Hydra is extended in a certain direction, after one or two minutes it contracts, bends over to a new position, then extends in a new direction. After about two minutes it contracts again, bends into a still different position, and again extends. This process is repeated at fairly regular intervals, so that after a time the Hydra has tried every position possible in its present place of attachment. This exploration of all parts of the surrounding region, of course, aids greatly in finding food.

Mast (1903) finds that when Hydra is heated from one side it does not move directly away from the source of heat, but merely moves in some random direction. In other words, the animal when heated merely tries a new position.

Mœbius (1873) describes the reaction of a large mollusk (Nassa) to chemical stimuli, as shown when a piece of meat is thrown into the aquarium containing them. They do not orient themselves in the lines of diffusion and travel directly toward the meat, but move " now to the right, now to the left, like a blind man who guides himself forward by trial with his stick. In this way they discover whether they are coming nearer or going farther away from the point from which the attractive stimulus arises" (*l. c.*, p. 9, translation). The reaction is thus a clear case of the method of trial and error. Experiments on the leech,

[*] By Mr. George Wagner.

by Miss Frances Dunbar, which I hope may soon be published, show that that animal finds its food in a similar manner. All searching is, of course, behavior on the plan of trial and error, and many organisms are known to search for food.

The righting reactions of organisms are among the most striking examples of trial and error in behavior. In the starfish, for example, when the animal is laid on its back " the tube feet of all the arms are stretched out and are moved hither and thither, as if feeling for something, and soon the tips of one or more arms turn over and touch the underlying surface with their ventral side " (Loeb, 1900, p. 62). As soon as these one or two arms have been successful, the others cease their efforts; the attached arms then turn the body over. If all the arms attempted to turn the animal at the same time, in other words, if there were no way of recognizing " success " in the trial, the animal could not right itself.

The righting reaction of the starfish shows much resemblance to the method, described above, by which a suspended Amœba passes to a solid. It is probable that a Difflugia, turned with the opening of the shell upward, would show a righting reaction essentially similar to that of the starfish.

The righting reaction of the flatworm Planaria, as described by Pearl (1903), is not so evidently brought about through the method of trial and error. Yet there are certain facts that indicate that this method is really essentially present here. Thus, Pearl shows that when the flatworm is prevented from righting itself in the usual way, its rights itself in another manner. Probably various reactions are tried; if the first does not succeed another may.

This peculiar form of the method of trial and error, in which several different reactions are tried even under but a single stimulus, is brought out by Mast (1903) in the behavior of Planaria under other conditions. If the water containing the flatworms is heated, the animals give, as the temperature rises, practically all the reactions that they ever give under any conditions.

We have thus, as the temperature rises and the stimulation increases, the following reactions given consecutively: positive, negative, crawling, righting, and final (all the reactions described by Pearl, with the exception of the food reactions, and the final reaction in addition). (Mast, 1903, p. 185.)

We shall have occasion to inquire as to the significance of this responding to the same stimulus by many different reactions when we take up, in another connection, certain similar phenomena in Stentor.

In the higher vertebrates, as we have mentioned at the beginning, the method of trial and error plays a very large part. It is here especially that it has been recognized as a definite type of behavior in the work of Lloyd Morgan, Thorndike, and others. With the details of its manifestations in higher animals we need not concern ourselves here.

Let us now return to the method of trial and error in the infusoria. Here it is most strongly marked, and certain general problems which arise from it are sharply defined. It is possible to formulate the reaction method of the infusoria as follows: When effectively stimulated by agents of almost any sort the organism moves backward, and turns toward a structurally defined side X, while at the same time it may continue to revolve on its long axis. As thus stated, the reaction method seems exceedingly simple and stereotyped, and as such I presented the behavior of these organisms in a former general paper.* If we limit ourselves to a consideration of the reaction itself, this seems inevitable, although certain additional reactions have been described since that paper was written. But when we study the relation of this reaction method to the environmental conditions, the results are most remarkable, and a totally new set of problems appears. Whether the behavior is to be called simple and stereotyped, or complicated and flexible, is not so easy to decide. The relations of the reaction of the environmental conditions are, perhaps, the really essential point in animal behavior. What are the relations which we find in the organisms reacting in the way set forth above?

In general terms we find that through this reaction by trial and error the organisms are kept in conditions favorable to their existence, and prevented from entering unfavorable regions. Through it they keep out of hot and cold regions and collect in regions of moderate temperature. Through it they tend to keep out of strong or injurious chemicals and out of regions where the osmotic pressure is much above or below that to which they are accustomed. Through it they gather in regions containing small amounts of certain chemicals, not leaving them for regions where there is either more or less of these chemicals. When oxygen is needed they collect through this reaction in regions containing oxygen; when the oxygen pressure is high, they do not react with reference to oxygen, or through this reaction they avoid regions containing much oxygen. Through this reaction organisms which contain chlorophyll, and therefore need light, gather in lighted regions or move toward the source of light; through the same reaction the same organisms avoid very powerful light.† In all these cases,

---

* Psychology of a Protozoan. Amer. Journ. Psychol., vol. 10, 1899, pp. 503–515.

† For details, see the author's Studies, already referred to, and the preceding contributions. In papers by Engelmann (1882, a) and Rothert (1901) the reaction method involved is also described for certain organisms, though these writers, like the present author in his earlier papers, did not bring out the relations to a general method of trial and error. Engelmann, however, characterized the reaction of Euglena to light directly as a "Probiren"—a "trial."

when there is error the organism goes back and tries a new direction, or a whole series of new directions.

But what constitutes "error"? This is a fundamental question for this method of behavior. Why does the organism react to some things by turning away and trying new directions, to others not? Why do they react thus on coming to certain chemicals, and on leaving others? Why do they react thus on coming to a strong chemical, and also on leaving a weak solution of the same chemical? Why does the same organism react thus to strong light, and also to darkness? To heat and also to cold? What decides whether a certain condition is "error" or not? A list of all the different agents that must be considered "error" from the standpoint of this reaction method reveals, so far as chemical or physical classification is concerned, a most heterogeneous and even contradictory collection. What is the common factor which makes them all error?

Examination shows that error from the standpoint of this behavior is as a rule *error* also from the standpoint of the general interests of the organism, considering as the interests of the organism the performance of its normal functions, the preservation of its existence, and the production of posterity. In general the organism reacts as error to those things which are injurious to it, while in those conditions which are beneficial it continues its normal activities. There are some exceptions to this, but in a general view it is clearly evident. There is no common thread running through all the different agents which constitute " error" in the reactions, save this one, that they *are* error from the standpoint of the general interests of the organism.

How can we account for the fact that these lowest organisms react to all sorts of things that are injurious to them by a reaction which tends to remove them from the action of the agent,—by a negative reaction? The first response to this question must be another question. How can we account for the fact that in man we have the same condition of affairs? How does it happen that *we* respond by drawing back both from flame and from ice, though these act physically in opposite ways? Why do we seek light, but avoid a blinding glare? Why do we receive without opposition certain chemical stimuli (odors and tastes) and avoid others? The facts are quite parallel in man and in the lowest organisms in these respects. In man certain stimuli cause reactions which tend to remove the organism from the source of the stimulus (negative reactions), while others have the opposite effect; this is true also of Euglena and Paramecium. In both cases the stimuli which produce the negative reaction form a heterogeneous collection from the chemical or physical standpoint. In both cases the stimuli producing the negative reaction are in general injurious to the organism. The problem is one for the highest and for the lowest organisms.

In ourselves the stimuli which induce the negative reaction bring about the subjective state known as *pain* (in varying degrees, from discomfort or dislike to anguish), and popularly we consider that the drawing back is due to the pain. Is there ground for this view? Or is the reaction entirely accounted for by the chemical and physical processes involved? When the burnt child draws its hand back from the flame, does the state of consciousness called pain have anything to do with the reaction?

Without attempting to answer this question, we wish to point out the bearing of possible answers on our problem in the lower organisms. If we hold that in man we cannot account for the reaction without taking into consideration the pain, then we must hold to the same view for the lower organisms. If we maintain that in man we cannot account for the selection of such a heterogeneous group of conditions for the negative response—conditions seeming to have nothing in common save that they cause pain—without taking into consideration the pain, then we are forced to the same conclusion in the unicellullar organisms, for here we have a precisely parallel series of phenomena. Anyone, then, who holds that pain is a necessary link in the chain of events in man must consider that we are undertaking a hopeless task in trying to account for the reactions of the lower organisms purely from the chemical and physical conditions. And the converse is also true. Anyone who holds that we can account fully for the reactions of Euglena or Paramecium, purely from the physico-chemical conditions, without taking into account any states of consciousness, must logically hold that we can do the same in man. The method of trial and error implies some way of distinguishing error; the problem is: How is this done? The problem is one, so far as objective evidence goes, throughout the animal series.

We can, of course, know nothing of pain in any organism except the self, and we can, in a purely formal way at least, solve our problem equally well (or equally ill) without taking pain directly into consideration. Even in man we must hold that pain is preceded or accompanied in every case by a certain physiological condition. And if there is something common to all states of pain, it would appear that there must be something common to all the physiological states accompanying or preceding pain. We thus get a common basis for all the negative reactions; if they are preceded or accompanied by a common physiological state, this state will serve formally as an explanation for the common reaction, fully as well as would a common state of consciousness. The facts could be formulated as follows: In any animal, from the lowest up to man, a certain heterogeneous set of agents, which are in general injurious, produce a certain physiological state, common to all; as a consequence or concomitant of this physiological state a negative reaction follows. In man this physiological state is accompanied by pain. The common physiological state might then properly receive a name which brings out its relation to the state accompanied in man by pain, for example, J. Mark Baldwin's "organic analogue of pain."

We have left under this formulation the fundamental question as to how a set of agencies that are quite heterogeneous from a chemical or physical standpoint can produce a common physiological state. This question is, of course, of precisely the same order of difficulty as that which asks how the same heterogeneous set of agents can produce a common state of consciousness, namely, pain. We therefore lose nothing, so far as this problem is concerned, by substituting "a common physiological state" for "pain" in dealing with the subject. But to attempt to deal with the problem of negative reactions in the lower organisms without recognizing that they are conditioned in the same way as the negative reactions of man—without admitting the existence of some physiological state analogous to that which is accompanied by pain in man, is, I believe, to close one's eyes to patent realities.

We have seen above that the method of trial and error involves some way of distinguishing error. But do not some of the facts indicate that it involves, at least sometimes, also some way of distinguishing the opposite of error; that is, what we may call success? For most of the reactions of the infusoria this seems not necessary, for what the organism does when successful is merely to continue the condition in which it finds itself at the time. There is then no objective evidence that a stimulus is acting at this time. In these organisms it seems to be chiefly the injurious or negative stimuli that induce a motor reaction. But consider the floating Amœba, which sends forth pseudopodia in all directions. Finally one of these pseudopodia comes in contact with a solid, and to this stimulus the Amœba reacts positively. Now all the other pseudopodia, though the external conditions directly affecting them remain the same, become retracted, and the whole Amœba moves toward the pseudopodium in contact. This withdrawal of the other pseudopodia requires for its explanation a change in physiological state which can be due only to the success of the pseudopodium that has come in contact with a solid. There is certainly no basis here for considering the reaction as due to an "organic analogue of pain"; possibly a case could be made out, on the other hand, for a physiological state corresponding to that which conditions *pleasure* in ourselves. Possibly similar considerations hold for the positive reactions of other organisms—infusoria, etc.—to solids.

There appears to be a similar state of affairs in the righting reaction.

When the starfish is placed on its back, the physiological state existing induces all the arms to initiate feeling movements; the animal tries to reach a solid with its tube feet. As soon as one or two arms have succeeded, this success is recognized by the cessation of effort on the part of the other arms. Their physiological state has changed to one corresponding to success.

We may sum up our discussion on these points as follows: The method of trial and error involves some way of distinguishing error, and also, in some cases at least, some method of distinguishing success. The problem as to how this is done is the same for man and for the infusorian. We are compelled to postulate throughout the series certain physiological states to account for the negative reactions under error, and the positive reactions under success. In man these physiological states are those conditioning pain and pleasure.

The "method of trial and error," as this phrase is used in the present paper, is evidently the same as reaction by "selection of overproduced movements," which plays so large a part in the theories of Spencer and Bain and especially in the recent discussions of behavior by J. Mark Baldwin. To this aspect of the matter the present writer will return in the future.

This method of trial and error, which forms the most essential feature of the behavior of these lower organisms, is in complete contrast with the tropism schema, which has long been supposed to express the essential characteristics of their behavior. The tropism was conceived as a fixed way of acting, forced upon the organism by the direct action of external agents upon its motor organs. Each class of external agents had its corresponding tropism; under its action the organism performed certain forced movements, usually resulting in its taking up a rigid position with reference to the direction from which the stimulus came. Whether it then moved toward or away from the source of stimulus was determined by accidental conditions, and played no essential part in the reaction. There was no trial of the conditions; no indication of anything like what we call choice in the higher organisms; the behavior was stereotyped. Doubtless such methods of reaction do exist. In the reactions of infusoria to the electric current (an agent with which they never come into relation in nature), there are certain features which fit the tropism schema, and in the instincts—the "Triebe"—of animals there are features of this stereotyped character. The behavior of animals is woven of elements of the most diverse kind. But certainly in the lower organisms which we have taken chiefly into consideration the behavior is not typically of the stereotyped character expressed in the tropism schema. The method of trial and error is flexible; indeed, plasticity is its essential characteristic. Working in the lowest organisms with very simple factors, it is nevertheless capable of development; it leads upward. The tropism leads nowhere; it is a fixed, final thing, like a crystal. The method of trial and error on the other hand has been called the "method of intelligence" (Lloyd Morgan, 1900, p. 139); it involves in almost every movement an activity such as we call choice in higher organisms. With the acquirement of a *finer perception of differences* the organism acting on the method of trial and error rises at once to a higher grade in behavior. Combining this with the development of sense organs and the differentiation of motor apparatus, the path of advancement is wide open before it.

The most important step in advance is that shown when the results of one reaction by trial and error become the basis for a succeeding reaction. The method of reacting which leads to success is determined by trial; after it is once or several times thus determined, the trials are omitted, and the organism at once performs the successful action. This is *intelligence*, according to Lloyd Morgan (1900, p. 138), and it is as leading to this result that the method of trial and error can be characterized also as the method of intelligence. In intelligent action, while the organism must react the first time by the method of trial and error, it need not begin all over again each time the same circumstances are presented. Do we find any indication of such action among unicellular organisms?

In Stentor we find action of this character to a certain extent. It does not continue reacting strongly to a stimulus that is not injurious, but after a time, when such a stimulus is repeated, it ceases to react, or reacts in some less pronounced way than at first. To an injurious stimulus, on the other hand, it does continue to react, but not throughout in the same manner. When such a stimulus is repeated, Stentor tries various different ways of reacting to it. If the result of reacting by bending to one side is not success, it tries reversing the ciliary current, then contracting into its tube, then leaving its tube, etc. (details in Jennings, 1902, *a*). This is clearly the method of trial and error passing into the method of intelligence, but the intelligence lasts for only very short periods. To really modify the life of the organism in any permanent way, as happens in higher animals, the method of reacting discovered to be successful by the method of trial and error should persist for a long time. Apparently this is not the case for unicellular organisms, but further work is needed on this point.

An application of the method of trial and error similar to that of Stentor is found under certain circumstances in the flatworm. Pearl (1903) found that after the animal had reacted to a repeated mechanical stimulus for a long time by turning away from it, it suddenly reversed the reaction, and turned far toward the side on which the stimulus was acting. Mast (1903) showed, as we have seen, that when the flatworm is heated it tries successively almost every form of reaction which it has at command. Such results have a most important bearing on the problem of the relation of the reaction method to the stimulus. Neither direct action of the stimulus on the motor organs as separate entities, nor a typical fixed interconnection of sense organs and motor organs can explain such results. As a result of continued strong stimulation the organism passes from one physiological state to another, and each physiological state has its concomitant method of reaction.

The present paper may be considered as the summing up of the general results of several years' work by the author on the behavior of the lowest organisms. This work has shown that in these creatures the behavior is not as a rule on the tropism plan—a set, forced method of reacting to each particular agent—but takes place in a much more flexible, less directly machine-like way, by the method of trial and error. This method involves many of the fundamental qualities which we find in the behavior of higher animals, yet with the simplest possible basis in ways of action; a great portion of the behavior consisting often of but one or two definite movements, movements that are stereotyped when considered by themselves, but not stereotyped in their relation to the environment. This method leads upward, offering at every point opportunity for development, and showing even in the unicellular organisms what must be considered the beginnings of intelligence* and of many other qualities found in higher animals. Tropic action doubtless occurs, but the main basis of behavior is in these organisms the method of trial and error.

---

*Throughout this paper a number of terms are used whose significance as they are commonly employed is determined by our subjective experience. But all these terms (save those directly characterized as "subjective states," or "states of consciousness") will be found susceptible also of definition from certain objective manifestations, and it is in this objective sense that they are used in the present paper. Thus "perception" of a stimulus means merely that the organism reacts to it in some way; "discrimination" of two stimuli means that the organism reacts differently to them; "intelligence" is defined by the objective manifestations mentioned in the text, etc. These terms are employed because it would involve endless circumlocution to avoid them; they are the vocabulary that has been developed for describing the behavior of men, and if we reject them, it is almost impossible to describe behavior intelligibly. When their objective significance is kept in mind there is no theoretical objection to them, and they have the advantage that they bring out the identity of the objective factors in the behavior of animals with the objective factors in the behavior of man.

LITERATURE CITED.

BALDWIN, J. MARK.
   1897. Mental development in the child and in the race. Methods and processes. Second Edition. 496 pp. New York.

BERNSTEIN, J.
   1900. Chemotropische Bewegungen eines Quecksilbertropfens. Arch. f. d. ges. Physiol., Bd. 80, pp. 628-637.

BERTHOLD, G.
   1886. Studien über Protoplasmamechanik. 332 pp., 7 pl. Leipzig.

BLOCHMANN, F.
   1894. Kleine Mitteilungen über Protozoen. Biol. Centralb., Bd. 14, pp. 82-91.

BÜTSCHLI, O.
   1878. Beiträge zur Kenntniss der Flagellaten und einiger verwandten Organismen. Zeitschr. f. wiss. Zool., Bd. 30. pp. 205-281, pls. 11-15.
   1880. Protozoa, I Abth. Bronn's Klassen und Ordnungen des Thierreichs. Leipzig & Heidelberg.
   1892. Untersuchungen über mikroskopische Schäume und das Protoplasma. 234 pp., 6 pl. Leipzig.

CARTER, H. J.
   1863. On *Amœba princeps* and its reproductive cells, compared with *Æthalium, Pythium, Mucor,* and *Achlya*. Ann. and Mag. of Nat. Hist. (3). Vol. 12, pp. 30-52, pl. 3.

CLAPARÈDE, ED., ET LACHMANN, J.
   1858-'9. Études sur les infusoires et les rhizopodes. T. I. 482 pp., 24 pl. Genève.

DAVENPORT, C. B.
   1897. Experimental Morphology. Vol. I. 280 pp. New York.

ENGELMANN, T. W.
   1882. Bacterium photometricum, ein Beitrag zur vergleichenden Physiologie des Licht- und Farbensinnes. Arch. f. d. ges. Physiol., Bd. 30, pp. 95-124. Pl. 1.
   1882 a. Ueber Licht- und Farbenperception niederster Organismen. Arch. f. d. ges. Physiol., Bd. 29, pp. 387-400.

GARREY, W. E.
   1900. The effect of ions upon the aggregation of flagellated infusoria. Amer. Journ. Physiol., Vol. 3, pp. 291-315.

GROOM, T. T., & LOEB, J.
   1890. Der Heliotropismus der Nauplien von Balanus perforatus und die periodischen Tiefenwanderungen pelagischer Tiere. Biol. Centralb., Bd. 10, pp. 160-177.

HARRINGTON, N. R., and LEAMING, E.
   1900. The reaction of Amœba to light of different colors. Amer. Journ. of Physiol., Bd. 3, pp. 9-18.

HODGE, C. F., & AIKINS, H. A.
   1895. The daily life of a Protozoan; a study in comparative psychophysiology. Amer. Journ. Psychology, Vol. 6, pp. 524-533.

HOLMES, S. J.
   1901. Phototaxis in the Amphipods. Amer. Journ. Physiol., Vol. 5, pp. 211-234.
   1903. Phototaxis in Volvox. Biol. Bulletin, Vol. 4, pp. 319-326.

HOLT, E. B., and LEE, F. S.
   1901. The theory of phototactic response. Amer. Journ. Physiol., Vol. 4, pp. 460-481.

JENNINGS, H. S.
   1897. Studies on reactions to stimuli in unicellular organisms. I. Reactions to chemical, osmotic, and mechanical stimuli in the ciliate infusoria. Jour. of Physiol., Vol. 21, pp. 258-322.

JENNINGS, H. S.
1899. Studies, etc. II. The mechanism of the motor reactions of Paramecium. Amer. Journ. Physiol., Vol. 2, pp. 311-341.
1899 a. Studies, etc. IV. Laws of chemotaxis in Paramecium. Amer. Journ. Physiol., Vol. 2, pp. 355-379.
1900. Studies, etc. V. On the movements and reactions of the Flagellata and Ciliata. Amer. Journ. Physiol., Vol. 3, pp. 229-260.
1900 a. Reactions of infusoria to chemicals. A criticism. Amer. Natural., Vol. 34, pp. 259-265.
1901. On the significance of the spiral swimming of organisms. Amer. Naturalist, Vol. 35, pp. 369-378.
1902. Artificial imitations of protoplasmic activities and methods of demonstrating them. Journ. of Appl. Microsc. and Lab. Methods, Vol. 5, pp. 1597-1602.
1902 a. Studies, etc. IX. On the behavior of fixed infusoria (*Stentor* and *Vorticella*), with special reference to the modifiability of protozoan reactions. Amer. Journ. Physiol., Vol. 8, pp. 23-60.
1903. Rotatoria of America. II. A monograph of the *Rattulidæ*. Bull. U. S. Fish Commission for 1902, pp. 273-352.

JENNINGS, H. S., and CROSBY, J. H.
1901. Studies, etc. VII. The manner in which bacteria react to stimuli, especially to chemical stimuli. Amer. Journ. Physiol., Vol. 6, pp. 31-37.

JENNINGS, H. S., and JAMIESON, CLARA.
1902. Studies, etc. X. The movements and reactions of pieces of ciliate infusoria. Biol. Bull., Vol. 3, pp. 225-334.

JENSEN, P.
1896. Ueber individuelle physiologische Unterschiede zwischen Zellen der gleichen Art. Arch. f. d. ges. Physiol., Bd. 62, pp. 172-200.
1900. Ueber den Aggregatzustand des Muskels und der lebendigen Substanz überhaupt. Arch. f. d. ges. Physiol., Bd. 80, pp. 176-228.
1901. Untersuchungen über Protoplasmamechanik. Arch. f. d. ges. Physiol., Bd. 87, pp. 361-417.
1902. Die Protoplasmabewegung. Asher & Spiro's Ergebnisse der Physiol., erster Jahrgang. 42 pp.

KÜHNE, W.
1864. Untersuchungen über das Protoplasma und die Contractilität. 158 pp., 8 pl. Leipzig.

LE DANTEC, F.
1894. Études biologiques comparatives sur les Rhizopodes lobés et réticulés d'eau douce. Bull. scient. de la France et de la Belgique. T. 26, pp. 56-99.
1895. Sur l'adhérence des Amibes aux corps solides. C. R. Acad. Sci., Paris. T. 120, pp. 210-213.

LEIDY, J.
1879. Freshwater Rhizopods of North America. Report of the U. S. Geol. Survey of the Territories. Vol. 12, 324 pp., 48 pl.

LOEB, J.
1893. Ueber künstliche Umwandlung positiv heliotropischer Thiere in negativ und umgekehrt. Arch. f. d. ges. Physiol., Bd. 54, pp. 81-107.
1897. Zur Theorie der physiologischen Licht- und Schwerkraftwirkungen. Arch. f. d. ges. Physiol., Bd. 66, pp. 439-466.
1900. Comparative physiology of the brain and comparative psychology. 309 pp. New York.

MAST, S. O.
1903. Reactions to temperature changes in Spirillum, Hydra, and fresh-water Planarians. Amer. Journ. Physiol., Vol. 10, pp. 165-190.

MENDELSSOHN, MAURICE.
1902 a. Recherches sur la thermotaxie des organismes unicellulaires. Journ. de Physiol. et de Path. gen., T. 4, pp. 393-410.
1902 b. Recherches sur l'interférence de la thermotaxie avec d'autres tactismes et sur le mecanisme du mouvement thermotactique. *Ibid.*, pp. 475-488.
1902 c. Quelques considérations sur la nature et la role biologique de la thermotaxie. *Ibid.*, pp. 489-496.

MOEBIUS, K.
1873. Die Bewegungen der Thiere und ihr psychischer Horizont. 20 pp. Kiel.
MOORE, ANNE.
1903. Some facts concerning the geotropic gatherings of Paramecia. Amer. Journ. Physiol., Vol. 9, pp. 238-244.
MORGAN, C. LLOYD.
1894. Introduction to Comparative Psychology. 382 pp. London.
1900. Animal Behavior. 344 pp. London.
NAEGELI, C.
1860. Ortsbewegungen der Pflanzenzellen und ihren Theile (Strömungen). Beiträge zur wissenschaftlichen Botanik, Heft 2, pp. 59-108.
NAGEL, W. A.
1894. Beobachtungen über den Lichtsinn augenloser Muscheln. Biol. Centralb., Bd. 14, pp. 385-390.
OSTWALD, WILHELM.
1902. Vorlesungen über Naturphilosophie. 457 pp. Leipzig.
OSTWALD, WOLFGANG.
1903. Zur Theorie der Richtungsbewegungen schwimmender niederer Organismen. Arch. f. d. ges. Physiol., Bd. 95, pp. 23-65.
PEARL, R.
1900. Studies on electrotaxis. I. On the reactions of certain infusoria to the electric current. Amer. Journ. Physiol., Vol. 4, pp. 96-123.
1903. The movements and reactions of fresh-water Planarians. A study in animal behavior. Quart. Journ. Micr. Sci., Vol. 46, pp. 509-714.
PENARD, EUG.
1890. Études sur les Rhizopodes d'eau douce. Mém. de la Soc. de Physique et d'Histoire naturelle de Genève, T. 31, pp. 1-230. Pls. 1-11.
1902. Faune rhizopodique du bassin du Leman. 714 pp. Genève.
PROWAZEK, S.
1900. Protozoenstudien, II. Arb. a. d. zool. Inst. Wien, Bd. 12, pp. 243-297, 298-300.
1901. Beiträge zur Protoplasma-Physiologie. Biol. Centralb., Bd. 21, pp. 87-95.
PÜTTER, AUG.
1900. Studien über Thigmotaxis bei Protisten. Arch. f. Anat. u. Physiol., physiol. Abth., Supplementband, 1900, pp. 243-302.
QUINCKE, G.
1888. Ueber periodische Ausbreitung von Flüssigkeitsoberflächen und dadurch hervorgerufene Bewegungserscheinungen. Sitzb. d. kgl. preuss. Akad. d. Wiss. zu Berlin, Bd. 34, pp. 791-804.
RADL, EM.
1903. Untersuchungen über den Phototropismus der Tiere. Leipzig. 188 pp.
ROESLE, E.
1902. Die Reaktion einiger Infusorien auf einzelne Induktionsschläge. Zeitschr. f. allg. Physiol., Bd. 2, pp. 138-168.
ROTHERT, W.
1901. Beobachtungen und Betrachtungen über tactische Reizerscheinungen. Flora, Bd. 88, pp. 371-421.
ROUX, W.
1891. Ueber die "morphologische Polarisation" von Eiern und Embryonen durch den electrischen Strom. Sitz.-Ber. d. k. Akad. d. Wissensch. zu Wien, Math. u. Naturw. Classe, Bd. 101, pp. 27-228 (Ges. Abhdlg., Bd. 11, pp. 540-765).
RHUMBLER, L.
1898. Physikalische Analyse von Lebenserscheinungen der Zelle. I. Bewegung, Nahrungsaufnahme, Defäkation, Vacuolen-Pulsation, und Gehäusebau bei lobosen Rhizopoden. Arch. f. Entw.-mech. der Organismen, Bd. 7, pp. 103-350.
1902. Der Aggregatzustand und die physikalischen Besonderheiten des lebenden Zellinhaltes. Zeitschr. f. allg. Physiol., Bd. 1, pp. 279-388; Bd. 2, pp. 183-340.
SCHULZE, F. E.
1875. Rhizopodenstudien. IV. Arch. f. Mikr. Anat., Bd. 11, pp. 329-353, pls. 18-19.

SOSNOWSKI, J.
    1899. Untersuchungen über die Veränderungen der Geotropismus bei Paramecium aurelia. Bull. Internat. de l'Acad. d. Sci. de Cracovie, Mars, 1899, pp. 130-136.
STAHL, E.
    1884. Zur Biologie der Myxomyceten. Bot. Zeitung, Jhrg. 40, pp. 146-155; 162-175; 187-191.
STRASBURGER, E.
    1878. Wirkung des Lichtes und der Wärme auf Schwarmsporen. Jenaische Zeitschr., N. F., Bd. 12, pp. 551-625. Also separate, Jena. 75 pp. (The citations are to pages of the separate reprint.)
THORNDIKE, E. L.
    1898. Animal intelligence. An experimental study of the associative processes in animals. The Psychological Review, Monograph Supplement 2. 109 pp.
TOWLE, ELIZ. W.
    1900. A study in the heliotropism of Cypridopsis. Amer. Journ. Physiol., Vol. 3, pp. 345-365.
UEXKÜLL, J. v.
    1899. Die Physiologie der Pedicellarien. Zeitschr. f. Biol., Bd. 37, pp. 334-403.
    1900. Die Physiologie des Seeigelstachels. *Ibid.*, Bd. 39, pp. 73-112.
    1900 a. Die Wirkung von Licht und Schatten auf die Seeigel. *Ibid.*, Bd. 40, pp. 447-476.
    1903. Studien über den Tonus. I. Der biologische Bauplan von Sipunculus nudus. *Ibid.*, Bd. 44, pp. 269-344.
VERWORN, M.
    1889. Psycho-physiologische Protistenstudien. Experimentelle Untersuchungen. 219 pp., 6 pl. Jena.
    1890. Biologische Protistenstudien, II. Zeitschr. f. Wiss. Zool., Bd. 50, pp. 443-468, pl. 18.
    1890 a. Die polare Erregung der Protisten durch den galvanischen Strom. Arch. f. d. ges. Physiol., Bd. 46, pp. 281-303.
    1891. Die physiologische Bedeutung des Zellkernes. Arch. f. d. ges. Physiol., Bd. 51, pp. 1-118, pls. 1-6.
    1892. Die Bewegung der lebendigen Substanz. 103 pp. Jena.
    1895. Allgemeine Physiologie. Ein Grundriss der Lehre vom Leben. 584 pp. Jena.
    1897. Die polare Erregung der Protisten durch den galvanischen Strom. IV. Mittheilung. Arch. f. d. ges. Physiol., Bd. 65, pp. 47-62.
    1899. General Physiology. An Outline of the Science of Life. Transl. by F. S. Lee. New York. 615 pp.
WAGER, H.
    1900. On the eye-spot and flagellum of *Euglena viridis*. Journ. Lin. Soc. London, Zool., Vol. 27, pp. 463-481.
WALLENGREN, HANS.
    1902. Zur Kenntnis der Galvanotaxis. Zeitschr. f. allg. Physiol., Bd. 2, pp. 341-384.
    1903. Zur Kenntnis der Galvanotaxis. II. Eine Analyse der Galvanotaxis bei Spirostomum. *Ibid.*, Bd. 2, pp. 516-555.
WALLICH, G. C.
    1863 a. Further observations on an undescribed indigenous Amœba, with notes on remarkable forms of Actinophrys and Difflugia. Ann. and Mag. of Nat. Hist. (3), Vol. 11, pp. 365-371, pl. 9.
    1863 b. Further observations on the distinctive characters and reproductive phenomena of the amœban Rhizopods. Ann. and Mag. of Nat. Hist. (3), Vol. 12, pp. 329-337.
    1863 c. Further observations on the distinctive characters, habits, and reproductive phenomena of the amœban Rhizopods. Ann. and Mag. of Nat. Hist. (3), Vol. 12, pp. 448-468, pl. 8.
YERKES, R. M.
    1900. Reactions of Entomostraca to stimulation by light. II. Reactions of Daphnia and Cypris. Amer. Journ. Physiol., Vol. 4, pp. 405-422.